JANE'S POCKET BOOK OF SPACE EXPLORATION

ALSO AVAILABLE IN THIS SERIES:

JANE'S POCKET BOOK of AIRSHIPS
JANE'S POCKET BOOK of COMMERCIAL TRANSPORT AIRCRAFT
JANE'S POCKET BOOK of HOME-BUILT AIRCRAFT
JANE'S POCKET BOOK of LIGHT AIRCRAFT
JANE'S POCKET BOOK of MAJOR COMBAT AIRCRAFT
JANE'S POCKET BOOK of MAJOR WARSHIPS
JANE'S POCKET BOOK of MILITARY TRANSPORT and TRAINING AIRCRAFT
JANE'S POCKET BOOK of MISSILES
JANE'S POCKET BOOK of MODERN TANKS and ARMORED FIGHTING VEHICLES
JANE'S POCKET BOOK of NAVAL ARMAMENT
JANE'S POCKET BOOK of PISTOLS and SUB-MACHINE GUNS
JANE'S POCKET BOOK of RECORD-BREAKING AIRCRAFT
JANE'S POCKET BOOK of REMOTELY PILOTED VEHICLES
JANE'S POCKET BOOK of RESEARCH and EXPERIMENTAL AIRCRAFT
JANE'S POCKET BOOK of RIFLES and LIGHT MACHINE GUNS
JANE'S POCKET BOOK of SUBMARINES

JANE'S POCKET BOOK OF SPACE EXPLORATION

T. M. Wilding-White

COLLIER BOOKS
A Division of Macmillan Publishing Co., Inc.
New York

Macmillan Publishing Co., Inc.
866 Third Avenue, New York, N.Y. 10022
Collier Macmillan Canada, Ltd.

Library of Congress Catalog Card Number: 76-23520

First Collier Books Edition 1977

Second Printing 1978

Printed in the United States of America

INTRODUCTION

The story of Man's exploration of space has been short. Sputnik 1 was launched in October 1957 and only four years later America (and probably Russia, too) set out on a course to fulfil one of Man's oldest dreams — to walk upon another world. Now the conquest of the Moon is already history while Earth orbit has become both a hot commercial property and a cradle for a scientific renaissance. With routine reliability today's launch vehicles propel new spacecraft into orbit or interstellar trajectory. By January 1975 the grand total of payloads fired into space was 1,740. Of these, 1,007 had been destroyed re-entering our atmosphere or impacting on other worlds; but 684 were still circling the Earth while the remaining 43 were dispersing themselves throughout the solar system.

The conquest of space has emerged as a breathtaking surge of technological achievement, an adventure of brain rather than brawn. But the space age has now reached something of a watershed; behind lie nearly twenty years of enormous effort and soaring technical advance while ahead lie the beginnings of true cost-effective exploitation of space, to be signalled by the introduction in 1981 of a reusable space transporter, the space shuttle.

The story, however, is not just one of sensation and physical achievement. Throughout it has been characterised by revolutions in Earthbound capabilities and scientific knowledge. Application satellites serving communications, meteorology, navigation and Earth survey, are considered to have justified the entire space effort in financial terms alone. Space sciences and astronomy plus use of interplanetary probes promise the knowledge necessary for our survival, while Man himself in space is the key to future versatility in Earth orbit operations and perhaps in planetary exploration. This Pocket Book is designed as a detailed guide to current space programmes of all types. There is no such thing as a typical spacecraft; unlike aeroplanes, ships or tanks, there is little visual commonality and no conveniently delineated operational and production life. The rapidly changing and expanding scope of objectives and techniques throughout the first decades of space exploration has meant purpose-built hardware with production seldom exceeding a handful (with special exceptions such as reconnaissance satellites and certain launch vehicles). Each spacecraft, however, collects new data or serves an expanded role so that, even after the craft has died, its data are still being processed on Earth or applied to a successor. Until that successor appears, therefore, the spacecraft is still effectively current and valid.

In present-day terms a spacecraft is effectively a robot sent out to collect data and transmit it back to Earth. It consists of the main data-collecting instruments or implements, usually defined collectively as the payload, plus systems to protect and power it and to communicate with it. Protection normally means temperature control, a sophisticated feature demanding distribution of heating or cooling where required with minimal power consumption; radiation and meteoroid protection are also applied on some craft. Electrical power is usually very limited and supplied by rechargeable batteries, maintained by arrays of solar cells. Nuclear generators have, for needs of autonomy, entered service with the Jupiter Pioneers and the Viking Mars lander. Communications must be two-way. Instructions go to the craft which returns mission data, including TV pictures, plus telemetry on the condition or response of the subsystems themselves. Manned spacecraft, of course, also require voice channels plus life-support facilities, and usually optional manual control.

In addition there are the systems that position the payload — propulsion, navigation and attitude-control. Earth-orbiting craft often have no propulsion equipment at all unless orbital manoeuvring is required. Spacecraft destined for geo-stationary orbit invariably carry an "apogee" motor which, in effect, serves as an extra stage of the launcher in achieving final orbit positioning. Interplanetary craft need restartable engines for course-correction and, if required, insertion into the destination orbit. Navigation equipment is based on use of two or three different light-sensitive detectors usually graded to the Sun, the star Canopus and, if suitable, to Earth, to provide position fixes. Attitude-control and stabilisation systems ensure that the craft can be accurately orientated in space as required and held in position using small jets and/or gyros. Many craft are rotated in entirety to achieve gyroscopic rigidity, a technique known as spin-stabilisation. It is the limit of available attitude-control fuel that generally defines the operational life of any spacecraft. Finally, of course, all these systems must have their share of the protection, power supply and communication facilities.

To take the robot point a stage further, some recent craft have been equipped with their own brain. This has ranged from a simple automatic sequencer right up to sophisticated computers providing extensive control, trouble-shooting and correction facilities in addition to full implementation of the planned mission should command links from Earth fail.

The early spacecraft was essentially an integrated bundle of instruments and subsystems. Its design was challenging and its reliability low. Today it has evolved into a device which is highly sophisticated in terms of effectiveness and reliability of the systems yet greatly simplified by the standardisation of many of these systems. Most distinctive is the increasing use of a self-contained subsystems package or "bus", now appearing in modular form and heralding the day of the production-line spacecraft. Modular design will find another application when the space shuttle enters service with its ability to snatch satellites out of orbit and return them to Earth. Modular buses can then be fitted out with a new payload, refuelled and dispatched back to orbit again.

For simplicity the contents of this Pocket Book are divided into the five most widely used categories — manned, interplanetary, scientific and applications, plus launch vehicles. The section on manned spacecraft includes all types of craft ever flown because all are so familiar that even the earliest are still in the public domain. Likewise, a certain historical element exists among the interplanetary spacecraft selection by virtue of the distinctiveness of each step in Man's reaching out from his planet. Among the launch vehicles, too, are a few defunct but celebrated programmes. Soviet launchers are designated according to Charles S Sheldon's alphabetical/numerical system plus available Russian names. New Nasa-devised designators are also quoted in some cases.

The term scientific spacecraft covers a multitude of activities including engineering and technology-proving. Two other sub groups are most distinctive. The role of those craft designed to examine the radiation, content and forces in space within the neighbourhood of Earth is referred to as space sciences. Astronomy, the other group, applies to craft devoted to analysing the Sun and stars at all available levels of the electromagnetic spectrum.

The enormous potential of application satellites means that such programmes are beginning to diffuse into numerous specialised areas. Basically, however, there remain certain general groupings. The widest is communications which can now provide relay facilities for telephone, data, television, aeronautical and maritime control, linking points on a regional or global scale and operating through large distribution terminals or small community antennae for commercial, educational or military purposes. The permutations are endless and there is now a genuine shortage of sites free from mutual interference in geostationary orbit (circular/equatorial orbit at 22 300 miles distance where orbital speed is reduced to match that of the Earth's rotation and the satellite remains effectively fixed over a point on the surface). Then there are satellites for navigation, both civil and military, and a whole host of military satellites performing a variety of overt and secret tasks, notably visual and electronic reconnaissance, nuclear-test monitoring and missile early-warning. And then there are meteorology satellites which watch like Gods over the turmoils of our atmosphere and warn us of approaching vagaries of weather. Last and newest are Earth resources survey, a general category with unlimited specialised potential — mostly in the area of Earthbound interpretation of data rather than the collection of it.

The limitations on information available from the Soviet Union is inevitably reflected in an unbalanced coverage which in no way corresponds to levels of activity. For example, many of the separately detailed US Explorer-class scientific satellites have appropriate counterparts concealed within the so-called Cosmos scientific programme, a blanket label covering most civil and military activities and all failures. The Soviet launch rate consistently exceeds that of the USA — but again, this is no indication of the degree of return achieved.

Launch site and orbital data are generally excluded from the following profiles although approximate launch dates are provided to assist in locating particular missions. A detailed chronology for manned spaceflights and interplanetary probes will, however be found at the end of this Pocket Book.

TMW-W

Specialist operating agencies are included throughout in the following standard abbreviations:

CNES: Centre Nationale d'Etudes Spatiale
Comsat: Communications Satellite Corporation
DOD: Department of Defence
Eldo: European Launch vehicle Development Organisation (now replaced by ESA)
ESA: European Space Agency (for recent programmes)
Esro: European Space Research Organisation (now replaced by ESA)
ESSA: Environmental Science Services Administration
Intelsat: International Telecommunications Satellite Consortium
Intersputnik: Socialist bloc satellite communications consortium
Isro: Indian Space Research Organisation
MOD: Ministry of Defence
Nasa: National Aeronautics and Space Administration
Nasda: National Space Development Agency
NOAA: National Oceanic and Atmospheric Administration

(USSR) VOSTOK

Role: Manned spaceflight technology
Data: Vostok 6
Height: 24 ft 1 in (7.35 m). Without booster stage approx 15 ft (4.6 m)
Diameter: 8 ft 5 in (2.58 m). Capsule 7 ft 6 ins (2.3 m)
Launch weight: 10 370 lb (4 713 kg). Capsule about 5 280 lb (2 400 kg)
Spacecraft: Capsule is near-spherical re-entry module completely covered in ablative material and with antennae protruding forward. Pressurisation is Earth-type atmosphere. Capsule designed for automatic and ground control but with emergency manual controls. Circular hatch on side covers ejector seat on rails which operates automatically at 23 000 ft (7 000 m) so that crewman lands separately. Service module is roughly shaped like two truncated cones with bases together and with total length similar to 8 ft 5 in (2.58 m) max diameter. Upper end is ringed with spherical pressure vessels carrying oxygen and nitrogen. SM contains chemical batteries for 10 days' operation, attitude control system, and tanks of UDMH and nitric acid for restartable retro-rocket. Craft usually shown with launcher third stage still attached so that only upper part of SM is visible. Launcher stage is cylinder plus engine nozzle about 11 ft 6 in (3.5 m) high and quoted in overall spacecraft height although jettisoned after orbital insertion, allowing spacecraft to adopt slow tumble for temperature distribution
Payload: Single crew member was Valentina Tereshkova, first and only woman astronaut to date
Mission: Final Vostok mission, launched in July 1963 on three-day flight during which a "rendez-vous" approach to within 3 miles (5 km) of Vostok 5 was achieved. Completed 48 revolutions before landing on same day as Valery Bykovsky in Vostok 5 ended his five-day flight
Programme: First Soviet manned spaceflight programme initiated by five test flights, some carrying dogs, between May 1960 and March 1961 and designated Korabl Sputniks 1-5. First manned flight was Vostok 1 on April 12, 1961, which carried Yuri Gagarin on single orbit mission. Second flight in August and then two pairs — Vostok 3 and 4 in August 1962 and Vostoks 5 and 6 in July 1963 — attempted basic orbital rendez-vous by launch positioning. Vostok 5 remains longest solo spaceflight
Launch vehicle: A-1 (Vostok)

VOSKHOD

Role: Manned spaceflight technology
Data: Voskhod 2
Launch weight: 12 529 lb (5 683 kg)
Spacecraft: No details released but is believed to be modified Vostok. Capsule equipped with two seats but without ejector facility and therefore first and only manned programme not to have launch escape capability. Also equipped with small solid-propellant terminal-braking rockets to reduce parachute-controlled landing speed. Ejection hatch fitted with airlock of flexible material with second hatch fitted to outer end and extendible when required. Probably fitted externally and jettisoned before re-entry. Airlock pressurised at 6 psi (422 gm/cm^2) and cabin remained pressurised at normal Earth-type atmosphere throughout EVA activity
Payload: Crewmen Pavel Belyayev and Alexei Leonov. TV camera mounted externally to observe EVA
Mission: Voskhod 2 launched on March 18, 1965 and landed next day after 17 orbits. During second orbit, Leonov donned space suit and accomplished first space walk lasting 20 min, including 10 min free of craft. Failure of automatic landing system on 16th orbit forced extra revolution and manually controlled touchdown 1 400 miles (2 250 km) from target
Programme: Voskhod 2 was second and last manned variant on Vostok. Voskhod 1, launched in October 1964, had third crew couch instead of airlock and carried three crewmen without space suits or ejector seats on one-day prestige flight. Each mission preceded by unmanned test flights (Cosmos 47 and Cosmos 57). Final Voskhod mission flown as Cosmos 110 in February 1966. Craft carried two dogs on 22-day flight
Launch vehicle: A-2

Role: Manned spaceflight operations
Data: Soyuz 16
Height: Overall 23 ft 4 in (7.1 m). SM 7 ft 5 in (2.3 m). DM 7 ft 2 in (2.2 m). OM 8 ft 7 in (2.65 m)
Diameter: Array span 27 ft 3 in (8.3 m). SM 9 ft 0 in (2.7 m). DM 7 ft 3 in (2.2 m). OM 7 ft 3 in (2.2 m)
Launch weight: Overall 14 750 lb (6 690) kg). SM 5 850 lb (2 654 kg). DM 6 200 lb (2 812 kg). OM 2 700 lb (1 224 kg)
Spacecraft: Two/three-man spacecraft comprises service, descent and orbital modules (SM, DM, OM) arranged in tandem. SM is cylinder plus adapter skirt. Forward half pressurised for electronics and enclosed by radiator wall. Aft section contains four spherical propellant tanks plus four small pressurising tanks slotted into concave disc with engines mounted in centre. Single-nozzle primary engine flanked by twin-nozzle back-up engine, both with identical 882 lb (3.9 kN) thrust, plus four attitude-control thrusters of 29 lb (107 N) thrust each. Power (27 V) supplied by pair of solar arrays, each about 9 ft (2.7 m) long, deployed from opposite sides. DM is bowl-shaped with convex heat shield facing aft and attached to SM by open truss with external faring. Docking periscope projects from side. Main and reserve parachutes open at 29 000 ft (8 800 m). Heat shield jettisons just before touchdown to expose set of four solid-fuel terminal-braking rockets. Three crew couches arranged fanwise facing forward hatch into OM. Near-spherical OM contains equipment racks and working area plus entrance hatch on side and cylindrical docking-unit mount on nose. Attitude and docking control system includes ten more thrusters of 29 lb plus eight verniers of 3.3 lb (14.7 N) thrust. Pressurisation is Earth-type oxygen/nitrogen mix of 14.7 psi (1.033 gm/cm^2) adjustable to 10 psi (703 gm/cm^2) for ASTP docked phase. Cabin volume is 309 cu ft (8.3 m^3). Life-support supplies for up to 30 days can be carried. Launch-escape system lifts shroud plus OM and DM only and OM remains with shroud after DM separation
Payload: Crewmembers Anatoli Filpchenko, Nikolai Rukavishnikov
Mission: Six-day flight launched in December 1974 was dress rehearsal for ASTP mission Deployed separate docking unit to represent Apollo, made several mock docking manoeuvres and tested new pressurisation system
Programme: Total of 20 Soyuz-designated launches between April 1967 and July 1975 in wide range of applications. Soyuz 2 was unmanned target for Soyuz 3 rendez-vous. Soyuz 1, 3 and 4 had single-man crews; Soyuz 5, 7, 10 and 11 all three-manned; remainder two-manned. Soyuz 4 and 5 achieved first docking and transferred two crew by EVA. Soyuz 6, 7 and 8 achieved multiple rendez-vous. Soyuz 9 flew record 17-day mission. Soyuz 10, 11, 14, 15, 17, 18(A) and 18 all Salyut-related. Soyuz 19 was ASTP mission. Remainder as engineering flights. Early models had non-removable docking probe and 12 ft (3.6 m) long solar panels. Salyut ferry flights use batteries and have no solar arrays. Soyuz 1 and 11 missions resulted in death of crews. Soyuz 18(A) crew escaped in first ever manned launch abort in April 1975 (Soyuz mission number discounted). Unmanned Soyuz testing believed to include at least Cosmos 133, 140, 186/188, 212/213 (docking tests), 238, 379, 398, 414, 496, 573, 613, 638, 672 launched between November 1966 and August 1974. See also Zond. Unmanned supply-ferry version believed to be under test. Programme continuing
Launch vehicle: A-2

САЛЮТ
USSR

SALYUT

Role: Space station technology
Data: Salyut 3
Height: Approx 43 ft (12.9 m). With Soyuz 65 ft (20 m)
Diameter: 13 ft (4 m)
Launch weight: Approx 40 500 lb (18 400 kg). With Soyuz 55 000 lb (25 000 kg)
Spacecraft: Stepped cylinder has four sections. Forward end has passage or airlock compartment about 6 ft 6 in (2 m) in diameter, joined to docking unit and containing some experiments. Hatch leads aft into 10 ft (3.0 m) diameter forward section of 13 ft (4.0 m) diameter main chamber (external measurements) containing main control consols. Fitted to aft end is unpressurised cylindrical chamber containing Soyuz type propulsion system of two 880 lb (3.9 kN) thrust manoeuvring engines (one back-up) and four spherical tanks. Also contains attitude control system comprising four angled nozzles around main engine nozzles for yaw and pitch control with roll thrusters fitted to circumference of propulsion section (see Soyuz). Power supplied by three steerable solar arrays deployed in a "T" pattern from middle section. Docked Soyuz ferry Orbital Module provides sleeping accommodation and Soyuz is included in announced internal pressurised volume of 3 500 cu ft (100 m^3). Station electronics contained in sealed pressurised section at rear of largest compartment. Details based on Salut 1. Only overt difference is solar arrays — Salyut 1 had two pairs, one on each end section
Payload: Multipurpose payload includes Orion solar telescope and camera for Earth survey or reconaissance as well as medical, biological, processing and engineering experiments. Carries re-entry capsule. Payload and systems designed to operate automatically between crew visits
Mission: Third experimental orbital laboratory was launched in June 1974, operated until end December and re-entered atmosphere end January 1975. Received two-week visit from two-man Soyuz 14 crew in July 1974. Soyuz 15 failed to achieve rendez-vous in August and landed again after two days
Programme: Four Salyuts launched since April 1971. Salyut 2 disintegrated in orbit after launch in April 1973. Salyut 4 launched end December 1974. Cosmos 557 (May 1973) may have been test flight. Associated flights are Soyuz 10 and 11 (Salyut 1), Soyuz 14 and 15 (Salyut 3), Soyuz 17, 18A (abort) and 18 (Salyut 4). Crews boarded from Soyuz 11, 14, 17 and 18. Programme continuing
Launch vehicle: D-1-e

REMOVE BEFORE FLIGHT
McDONNELL

(USA — Nasa)

MERCURY

Role: Manned spaceflight technology
Data: Mercury-Atlas 9 (Faith 7)
Height: 9 ft 6 in (2.9 m). Escape tower 16 ft (4.9 m)
Diameter: 6 ft 2 in (1.88 m)
Orbital mass: 3 027 lb (1 376 kg)
Spacecraft: Minimum size spacecraft dictated by low-power launchers of the time is bell-shaped for thermal protection during re-entry. Titanium shell contains conical nickle alloy pressure vessel for crewman and all subsystems. Cylindrical nose section contains parachutes and radios. Ablative glass fibre heat shield extends 4 ft below craft after re-entry to cushion splashdown and act as sea-anchor. Jettisonable retro-package comprises 1 160 lb (5.16 kN)-thrust solid fuel rockets firing in sequence. 18-jet attitude control comprises duplicate sets of hydrazine jets of up to 24 lb (106.7 N) thrust plus mechanically controlled back-up system. Pressurisation is 100 per cent oxygen at 5.1 psi (359 gm/cm^2). 16.5 KW/h chemical-battery power supply. Escape tower has single, triple-nozzle, solid-propellant rocket of 50 000 lb (222.4 kN)-thrust plus 350 lb (1.5 kN)-thrust separation engine

Payload: Single crewman was Gordon Cooper. Several small operational experiments included release of a 6-in sphere with a flashing lights for research on rendezvous techniques, and ejection of tethered balloon for colour test
Mission: Last Mercury mission (and heaviest in series) was launched on May 15, 1963, and completed 22 orbits in 34 hr 20 min. Re-entry controlled manually after automatic controls failed and capsule splashed down within 4 miles (6.4 km) of recovery carrier
Programme: First US manned spaceflight programme comprised two sub programmes distinguished by launch vehicle — Mercury-Redstone (MR) and Mercury-Atlas (MA). MR involved four suborbital launches between December 1960 and September 1961, the first unmanned and the second carrying a chimpanzee plus two manned flights. MA involved three launch failures, one unmanned orbital shot and one chimpanzee in orbit before first of four manned flights (MA6-MA9) launched in February 1962. Planned 3-day MA 10 mission was cancelled. Prime contractor: McDonnell
Launch vehicle: Atlas D

GEMINI

Role: Manned spaceflight technology
Data: Gemini 12
Height: 18 ft 5 in (5.6 m). Re-entry module: 11 ft (3.35 m)
Diameter: (Base) 10 ft (3.0 m). Re-entry module: 3 ft 3 in (1.0 m) top, 7 ft 6 in (2.3 m) base
Orbital mass: 8 297 lb (3 763 kg)
Spacecraft: Conical spacecraft comprises re-entry module and adapter module. Re-entry module based on Mercury capsule but enlarged to provide 50 per cent increase in cabin volume. Cylindrical nose section has docking facilities and parachutes in forward compartment, duplicate sets of eight 25 lb (111 N)-thrust attitude-control thrusters and tanks for re-entry control in aft section. Cabin has side-by-side ejector seats with separate hatches instead of launch escape tower. Equipment not requiring pressurisation is fitted between titanium inner pressure hull and beryllium outer shell. Adapter module has two sections. "Retrograde" section contains four 2 500 lb (11.1 kN)-thrust solid-fuel braking rockets and set of docking control thrusters comprising two 85 lb (368 kN)-thrust deceleration jets and six 100 lb (445 kN)-thrust manoeuvring jets. Section jettisoned before re-entry. Equipment section at rear is jettisoned just before retro-burn and contains eight 25 lb (111 N)-thrust attitude-control jet, propellant tanks for both thruster systems plus fuel cells and hydrogen and oxygen supply. Fuel cells first introduced on Gemini 5
Mission: Gemini 12 launched on four-day mission with astronauts James Lovell and Edwin Aldrin on November 11, 1966. Docked with Agena target on third orbit. Aldrin performed one full EVA and two "stand-up" EVAs. Planned orbit raise using Agena engines cancelled because of engine perturbation
Programme: Series of 12 Geminis (first two unmanned) launched between April 1964 and November 1966 to test extended mission time, EVA activities and orbital manoeuvring as well as developing rendez-vous and docking techniques. First seven termed GT (Gemini-Titan) missions. GT 1 was unmanned orbital test and GT 2 was unmanned suborbital test. Last five all rendez-voused with Agena target vehicles and were designated GTA missions. Preceded by rendez-vous of Gemini 6 and 7. Gemini 7 achieved record 14-day flight and Gemini 11 used Agena engine to boost itself to record manned spaceflight apogee of 850 miles (1,368 km). Programme provided bridge between Mercury and Apollo. Gemini also planned for use as US Air Force Manned Orbiting Laboratory (MOL) but programme was cancelled after single test flight. Prime contractor: McDonnell
Launch vehicle: Titan II

Role: Manned lunar landing
Data: Apollo 11 Command/Service Module "Columbia"
Height: 34 ft 7 in (10.5 m). CM 12 ft 5 in (3.8 m). SM 24 ft 7 in (7.5 m)
Diameter: 12 ft 10 in (3.9 m)
Launch weight: 63 493 lb (28 800 kg). SM 51 243 lb (23 243 kg)
Spacecraft: Conical command module is stainless-steel honeycomb structure comprising three sections. Nose compartment contains parachutes and pair of negative pitch thrusters and encloses removable docking probe. Cabin, which is aluminium honeycomb pressure shell, has 210 cu ft (5.94 m^3) volume and three crew couches in level row facing forward. Aft section contains set of 10 attitude-control jets of 93 lb (413 N) thrust and hypergolic propellant tanks. CM is re-entry vehicle and has ablative heat shield of phenolic epoxy resin. Service module is aluminium honeycomb cylinder divided longitudinally into six compartments containing three fuel cells, their hydrogen and oxygen supply, duplicate attitude-control system comprising 16 jets of 100 lb (445 N) thrust, service propulsion system engine of 20 500 lb (911.2 kN) thrust plus hypergolic propellant tanks for SPS and control thrusters. SPS powers all manoeuvres, corrections, lunar orbit departure and retro-burns. Four-dish high-gain antenna projects from aft end of SM. Launch Escape Tower is open frame 33 ft (9.9 m) high, weighs 8,848 lb (4,000 kg) and has three solid-fuel motors — 147 000 lb (653.5 kN)-thrust main escape motor on top; 31 500 lb (1400 kN)-thrust tower-jettison motor; 2 400 lb (10.7 kN)-thrust escape pitch-control motor
Payload: Crewman Neil Armstrong, Michael Collins, Edwin Aldrin. Lunar module weighing 33 205 lb (15 061 kg) (detailed separately)
Mission: First lunar landing mission was launched on July 16 1969. After 1½ orbits, S-VB stage boosted CSM into translunar trajectory and 33 min later, CSM initiated transposition and docking with lunar module. On July 19, SPS firing of 5 min 57.5 sec placed CSM in initial lunar orbit of 69 x 190 mile (110 x 304 km). On July 20, Armstrong and Aldrin departed in LM "Eagle" for lunar landing while Collins remained orbiting in CSM "Columbia". Eagle redocked 26 hr 5 min later and was then jettisoned in lunar orbit. On July 21, after 59 hr in lunar orbit, SPS engine was fired for 2 min for trans-Earth injection. Only one mid-course correction required in each direction. Splashdown occurred on July 24 after flight lasting 195 hr 18 min 15 sec
Programme: Apollo II was first of six lunar landing missions (Apollos 11, 12, 14, 15, 16 and 17) plus one mission abort (Apollo 13) launched between July 1969 and December 1972. Final three were modified to extend capabilities and designated J-series (detailed separately). Preceded by four manned development flights, Apollos 7-10 (October 1968 to May 1969). Apollo 7 and 9 tested hardware in Earth orbit. Apollo 8 made first circumlunar flight and Apollo 10 made full approach rehearsal. Six "boilerplate" flights between February 1966 and April 1968 include one suborbital launch and one LM-only flight. Three final landing missions cancelled. CSM used subsequently for Earth orbit missions including three Skylab crew ferry flights in 1973 and Apollo-Soyuz Test Programme mission in 1975. Prime contractor: North American Rockwell
Launch vehicle: Saturn V

(USA — Nasa)

APOLLO LUNAR MODULE

Role: Manned lunar landing
Data: Apollo 11 Lunar Module (Eagle)
Height: 22 ft 11 in (7.0 m). Ascent stage 12 ft 4 in (3.76 m)
Diameter: 31 ft (9.45 m) across landing pads. Ascent stage 14 ft 1 in (4.29 m)
Launch weight: 33 205 lb (15 061 kg). Ascent stage 10 622 lb (4 818 kg)
Empty weight: 9 287 lb (4 212 kg). Ascent stage 4 804 lb (2 179 kg)
Spacecraft: Two-stage aluminium spacecraft comprises lower or descent stage and upper or ascent stage. Descent stage is unpressurised, octagonal structure with central compartment housing hypergolic-fuelled descent engine with gimballed nozzle. Engine throttleable between 1 050 lb and 9 850 lb (4.67 kN and 43.78 kN) thrust, or 10 to 65 per cent and at 95 per cent of 10 500 lb (46.67 kN) max thrust. Compartment enclosed by four square bays containing propellant and lunar surface experiment equipment and enclosing triangular bays containing engine controls and water, helium and oxygen tanks for environmental control. Landing legs have 37 in (0.94 m) diameter footpads from three of which extend 68 in (1.72 m)-long surface-sensing probes. Ascent stage has pressurised cabin 92 in (2.3 m) in diameter and 42 in (1.07 m) across, with 160 cu ft (4.5 m^3) habitable volume plus controls and instruments .32 in (0.8 m)-square lunar EVA exit hatch in front. Vertical tunnel at back leads to docking and transfer hatch. Mid-section compartment contains guidance computer and control servos plus hypergolic propellant tanks for ascent engine and control thrusters. Aft equipment bay fitted to midsection, contains batteries and electronics. Ascent engine fitted below midsection has fixed thrust of 3,500 lb (15.56 kN). Attitude control provided by four sets of four 100 lb (445 N)-thrust jets, including duplicate system, fitted to ascent stage. Power provided by four batteries in descent stage, two in ascent stage
Payload: Early Apollo Scientific Experiments Package (EASEP) was first automatic lunar surface station and comprised seismometer and laser reflector. Also carried solar wind collector foil, TV and still cameras. Crewmen Neil Armstrong and James Aldrin
Mission: Apollo 11 lunar module "Eagle" separated from CSM "Columbia" on July 20, 1969 in orbit of 62 x 70 miles (99 x 112 km) and then lowered own orbit with 30 sec burn to 9 x 62 miles (14.5 x 99 km). Powered descent burn lasted 12 min 36.3 sec to touchdown, 102 hr 45 min after launch, at 21;17;40 GMT. Surface stay of 21 hr 36 min included single EVA providing four man-hours and collection of 44 lb (19.9 kg) of samples. Ascent-stage burn of 7 min 15 sec secured orbit 11 x 55 miles (18 x 88 km) altitude and craft manoeuvred to rendez-vous with CSM in 57 x 72 mile (91 x 115 km) orbit
Programme: Lunar module had sole application as lunar landing vehicle. Also considered for modification into Apollo Telescope Mount on Skylab under early scheme. Apollo 13 LM served as emergency propulsion and life-support module after loss of Service Module. Development included one boilerplate flight and manned tests in Earth and lunar orbit (Apollo 5, 9 and 10). Prime contractor: Grumman
Launch vehicle: Saturn V/Apollo CSM

J-SERIES APOLLO

Role: Manned lunar exploration
Data: Apollo 17 Command/Service Module "America"
Height: 34 ft 7 in (10.0 m). CM 12 ft 5 in (3.8 m). SM 24 ft 7 in (7.5 m)
Diameter: 12 ft 10 in (3.93 m)
Launch weight: 66 844 lb (30 320 kg). SM 54 044 lb (24 514 kg)
Empty weight: SM 13 450 lb (6 100 kg)
Spacecraft: Similar to early Apollo but modified to provide considerably extended operational and scientific capability. Modifications already performed after Apollo 13 accident included fitting extra liquid oxygen and extra battery in SM. J-series required additional liquid hydrogen tank. All extras fitted within five of six longitudinal bays with remaining empty bay reserved for Scientific Instrument Module (SIM). SIM bay measures 9 ft 5 in x 5 ft (2.9 m x 1.5 m) with 1 700 lb (771 kg) payload capacity and has jettisonable cover panel. CM equipped with control panels for SIM and EVAs
Payload: Crewmen Eugene Cernan, Ronald Evans and Harrison Schmidt. SIM bay carries orbital survey package comprising lunar sounder, mapping camera, panoramic camera, laser altimeter, infra-red scanning radiometer, far ultra-violet spectrometer and solar and galactic u-v spectrometer. Also contains second Particles and Fields sub-satellite ejected into lunar orbit (detailed separately). Lunar module weighing 36 262 lb (16 448 kg) (detailed separately)
Mission: Final Apollo lunar landing mission was launched on December 7, 1972. First and only night launch was halted in final 30 sec for 2 hr. Lunar survey orbit was 62 x 80 miles (100 x 130 km). Remained in lunar orbit for record 75 revolutions lasting 147 hr 48 min. 67-min EVA by Evans during return flight to retrieve SIM films. Splashed down on December 19, 1972, completing longest Apollo mission of 301 hr 52 min with record sample collection of 257 lb (117 kg)
Programme: J-series Apollo comprised final three missions (Apollo 15, 16 and 17) launched between July 1971 and December 1972. Three further missions cancelled and their objectives largely reallocated among first three, including choice of SIM experiments and landing sites. Prime contractor: Rockwell
Launch vehicle: Saturn V

FOR EMERGENCY EGRESS ONLY
GRUMMAN

(USA — Nasa) J-SERIES APOLLO LUNAR MODULE

Role: Manned lunar landing
Data: Apollo 17 Lunar Module "Challenger"
Height: 22 ft 11 in (7.0 m). Ascent stage 12 ft 4 in (3.76 m)
Diameter: 31 ft (9.45 m) across landing pads. Ascent stage 14 ft 1 in (4.29 m)
Launch weight: 36 262 lb (16 448 kg)
Spacecraft: Basically similar to early Apollo lunar modules but extensively modified to increase payload capability and extend maximum lunar surface stay time from 35 hr to 78 hr. Additions: one 100 Ah battery raises power by 13 per cent to 796 Ah from total of five batteries, giving 80 h of max consumption; extra 27 lb (12 kg) oxygen tank increases supply by 50 per cent to 80 lb (36 kg) total; water supply nearly doubled to 780 lb (354 kg). Existing tanks and equipment rearranged with additional tanks in three of triangular equipment bays on descent stage. Remaining bay retained for Lunar Roving Vehicle. Descent stage propulsion system modifications include lengthening of nozzle by 10 in (25 cm) to increase specific impulse by 1.5 sec and increasing propellant supply by 6.3 per cent for longer burn time
Payload: Crewmen Eugene Cernan and Harrison Schmidt (first scientist-astronaut). Lunar Roving Vehicle (detailed separately) and fifth Apollo Lunar Surface Experiment Package (Alsep) automatic science station comprising heat flow experiment, seismic profile experiment, surface gravimeter, atmospheric composition analyser, ejecta and meteorite detector, surface electrical properties experiment and lunar neutron probe
Mission: Final lunar mission achieved landing at Taurus Littrow site on December 11, 1972 at 1955 GMT. Total surface stay time was record 75 hr including 44.2 manhours of EVA during three EVA periods. LRV traverses totalled 22 miles (35 km). Longest EVA period was 7 hr 37 min
Programme: Extended capability lunar module used for final three Apollo missions, designated J-series, (Apollos 15, 16 and 17). Provided facilities for scientific exploration in addition to technical achievement. Series limited by cancellation of three planned missions. Prime contractor: Grumman
Launch vehicle: Saturn V/Apollo CSM

LUNAR ROVING VEHICLE

Role: Lunar surface travel
Data: Lunar roving vehicle (Apollo 17)
Height: 3 ft 9 in (1.14 m)
Length: 10 ft 2 in (3.1 m). Wheelbase 7 ft 6 in (2.3 m)
Width: 6 ft 9 in (2.01 m). Track 6 ft (1.8 m)
Gross weight: 1 480 lb (672 kg). Empty weight 480 lb (218 kg)
Spacecraft: Apollo lunar exploration vehicle is aluminium structure with four wheels each 31 in (79 cm) in diameter and 9 in (23 cm) across and two crew seats. Wheels are wire mesh, each with its own electric motor in hub and together powered by two 36 V batteries (one back-up). Two and four-wheel steering options and inertial navigation facility. Max speed is 8.5 mph (14 km/h). Max range is 37 miles (60 km). Folds into package 3 ft x 5 ft x 5 ft 6 in (0.9 m x 1.5 m x 1.7 m) for stowage aboard lunar module
Payload: 1 000 lb (454 kg) allowance provides 400 lb (181 kg) for each of two crewmen and their equipment plus 200 lb (91 kg) for experiments, tools and samples. Included is tool rack of drills, probes, cameras and magnetometer carried behind seats. Lunar Communications Relay Unit which has high-gain dish antenna and low-gain mast antenna provides direct link with Earth when vehicle is out of range of lunar module. Colour TV camera is remotely controlled from Earth
Mission: Final Apollo mission involved three LRV excursions during total EVA time of 22 hr 5 min. Distance covered totalled 21 miles (33.5 km). Longest EVA lasted 7 hr 37 min with journey of 11.8 miles (18.9 km)
Programme: Lunar Roving Vehicle built for use with J-series Apollo lunar missions (Apollo 15, 16 and 17) and successfully used on all of them. Total of ten were built including test and training models. Prime contractor: Boeing
Launch vehicle: Saturn V/Apollo CSM/Lunar Module

(USA — Nasa)

SKYLAB

Role: Space station technology
Data: Skylab 1
Length: Overall 118 ft 5 in (36.1 m) in orbital configuration. OWS 48 ft 2 in (14.7 m). AM 17 ft 6 in (5.4 m). MDA 17 ft 4 in (5.3 m). ATM 13 ft 3 in (4.0 m)
Diameter: OWS 21 ft 6 in (6.6 m). AM 10 ft (3.5 m). MDA 10 ft (3.0 m). ATM 7 ft (2.15 m)
Launch weight: Overall 165 450 lb (75 048 kg). OWS 78 000 lb (35 380 kg). AM 49 000 lb (22 225 kg). MDA 13 800 lb (6 260 kg). ATM 24 650 lb (11 180 kg)
Spacecraft: US experimental space station comprises three sections in tandem, Orbital Workshop (OWS), Airlock Module (AM), Multiple Docking Adapter (MDA) plus Apollo Telescope Mount (ATM) swivelled after launch into 90° position to MDA. Cluster completed by docked Apollo ferry. OWS is converted Saturn IVB stage. Hydrogen tank is divided into two floors. Lower floor has crew quarters, medical experiments, exercise equipment and airlock to oxygen tank serving as a waste dump. Spacious domed upper floor is general storage and activity area and is fitted with two scientific airlocks. OWS enclosed in deployable meteoroid and thermal shield and carries two deployable 30 ft x 23 ft (9 m x 7 m) solar arrays generating 12.5 kW. AM has two sections. Aft section is 5 ft 5 in (1.65 m) in diameter, contains airlock and exit hatch and is part enclosed within OWS. Forward section 3 ft 9 in (1.2 m) long is 10 ft (3 m) in diameter and provides extension to MDA, containing main spacecraft controls. MDA has docking port at end and emergency port on side and carries thruster attitude control system. ATM is non-inhabitable experiment unit with four deployable solar arrays generating 10.5 kW. Total habitable volume is about 11 800 cu ft (334 m^3) not including Apollo ferry. Pressurisation uses 70 per cent oxygen, 30 per cent nitrogen at 5 psi (350 gm/cm^2), including Apollo CM during docked phase
Payload: Main experiments include ATM with six telescopes serving five study areas. MDA contains ATM control console plus Earth Resources Experiments Package of six cameras and sensors and a furnace for space manufacturing experiments
Mission: First US space station launched as Skylab 1 on May 14, 1973. Meteoroid shield and one OWS solar array torn off during launch jamming other array and allowing station to overheat. First crew erected sun shield, released array and completed 90 per cent of programme. Improved sunshield erected by second crew. Three-man crews completed stays of 28 days, 59 days and 84 days. EVAs made by all crews for repairs and ATM film changes totalled 41 hr 47 min including record 7 hr 1 min EVA by third crew. Skylab expected to remain in orbit until 1983-1986
Programme: Single space station programme involved launch of three Apollo CSM crew ferries designated Skylab 2, 3 and 4 on May 25, July 28 and November 16, 1973. Prime contractors: (ATM) Marshall Space Flight Centre (OWS and AM) McDonnell Douglas; (MDA) Martin Marietta
Launch vehicle: Two-stage Saturn V

EXIT

(USA/USSR) APOLLO-SOYUZ TEST PROGRAMME

Role: Manned spaceflight technology
Data: ASTP docking module
Height: 10 ft 4 in (3.15 m)
Diameter: 4 ft 8 in (1.4 m)
Launch weight: 4 436 lb (2 012 kg)
Spacecraft: Purpose-built unit for joint US-USSR flight is depressurisation chamber and is cylinder with two near-rectangular boxes on side, each containing two spherical tanks for gaseous oxygen and nitrogen supply. Structure is thick aluminium sheeting, permitted by surplus launcher capacity. Systems module/control panel occupies full-length segment internally and lockers and experiments are attached around wall. Apollo end has standard Lunar Module type docking drogue. Soyuz end is fitted with new docking unit identical to mating unit on Soyuz. Unit has docking ring with three guide leaves and enclosing interface seal. Attenuator arms extend ring 12 in (0.3 m) on active craft which makes capture with latches in guides. Retraction of ring impacts eight locking catches in interface seal. Crew adjustment time in DM is 30 min into Soyuz at 10 psi (703 gm/cm^2) oxygen/nitrogen and 40 min into Apollo at 5 psi (350 gm/cm^2) pure oxygen. Overall length of Apollo-Soyuz is 66 ft (20.0 m)
Payload: Apollo crewmen Thomas Stafford, Vance Brand, Donald Slayton. Soyuz crewmen Aleksey Leonov, Valeriy Kubasov. Total of 26 experiments includes nine in DM, one in Soyuz, rest in CSM. Two require no equipment. DM contains metallurgy furnace with seven related experiments plus Earth-environment experiment. Remainder includes three Earth environment, one Earth survey, four astonomy, six medical and two medical technology. Five experiments to be operated jointly during docked phase
Mission: Apollo launch followed Soyuz launch by 7.5 hr. DM launched with Apollo CSM and attached in orbit by transposition and docking manoeuvre. Docked with Soyuz two days later with Apollo/DM in active role. Craft separated after two days and Soyuz re-entered while Apollo continued on science mission for five days
Programme: Single mission, launched on July 15, 1975, was designed to test new androgenous docking unit and rendez-vous techniques and to prove international space rescue capability. Unit will be fitted to space shuttles which will have similar Earth-type pressurisation as Soviet craft and therefore docking module will not be required
Launch vehicle: Saturn 1B

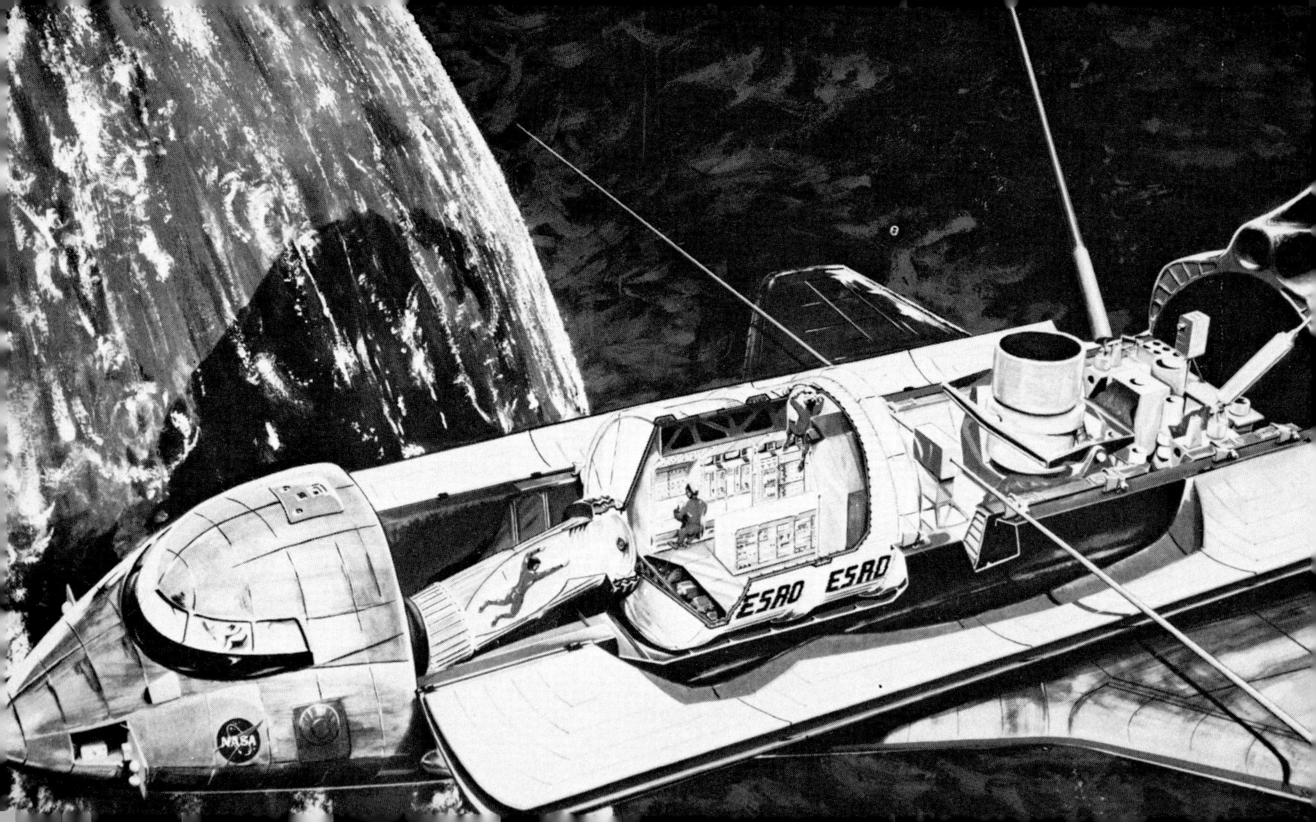
ESRO
ESRO
NASA

SPACELAB

Role: Manned space research
Data: Spacelab
Length: 49 ft (15 m) max without tunnel
Diameter: 14 ft (4.3 m)
Weight: About 29 600 lb (13 400 kg) with max payload
Spacecraft: Shuttle-borne space laboratory is re-usable modular design comprising pressurised laboratory module and equipment pallets. Lab has basic cylindrical unit containing standard housekeeping equipment plus similar optional extension piece for specialised experiments and controls. Each section is 9 ft (2.7 m) long and module is closed by slightly conical endpieces with central hatch either to access tunnel or to open pallets. Experiment airlock in top of each main section. Pallets are "U"-shaped structures with attachment fittings for large instruments. Each pallet is 10 ft (3 m) long and stressed for 3-ton loads. Up to three pallets can be used with full lab module or five for completely automatic payload. Spacelab is carried in 15 ft x 60 ft (4.6 m x 18.3 m) cargo bay of space shuttle (detailed separately). Flexible access tunnels, containing EVA airlock, links Spacelab with forward bulkhead of cargo bay. Shuttle provides thermal control, electric power, attitude control, communications and pressurisation of 20 per cent oxygen 80 per cent nitrogen at sea level pressure. Spacelab has back-up air recycling facility and own computer. Design life is 50 missions or five years
Payload: Up to four non-astronaut payload specialists can be carried by shuttle with three in Spacelab at any one time. Experiment hardware limits for various Spacelab configurations are about 12 100 lb (5 500 kg) for full lab module alone or short module plus 2-3 pallets; 20 000 lb (9 100 kg) for 2-3 pallets alone and 17 600 lb (8 000 kg) for five pallets alone. Pallet-only payloads are operable from within shuttle orbiter cabin or from Earth
Mission: Basic mission endurace defined by Shuttle capability of 7 days. Can be extended to 30 days by substituting supplies for payload
Programme: Spacelab, now under development by European consortium as Europe's contribution to Space Shuttle system, will be produced in limited quantities to bridge gap between Skylab and permanent space stations. First flight scheduled for 1980 as early Shuttle payload. Prime contractor: Erno/VFW-Fokker
Launch vehicle: Space Shuttle

USA
UNITED STATES OF AMERICA
USA

SPACE SHUTTLE ORBITER

Role: Re-usable space transport
Data: Space shuttle orbiter
Height: 57 ft (17.4 m)
Length: 122 ft (37.2 m)
Wingspan: 78 ft (23.8 m)
Weight: Empty 150 000 lb (68 040 kg). Landing 187 000 lb (84 800 kg)
Spacecraft: Re-usable space vehicle has conventional aluminium aircraft design with blended delta wings providing crossrange glide (departure from entry path) of 905 miles (1,460 km) and touchdown speed of 205 mph (346 km/h). No air-breathing engines carried. Most surfaces covered with low or high-temperature re-usable silica fibre insulation and nose and wing leading edge panelled with reinforced carbon carbon for re-entry thermal protection. Three Space Shuttle Main Engines, fuelled from jettisonable external propellant tank, provide launch propulsion (see Space Shuttle Booster). Orbital manoeuvring system (OMS) comprises two 6 000 lb (26.7 kN)-thrust engines for orbit insertion, manoeuvring and re-entry braking. Each engine mounted with own 5,970 lb (2,708 kg) supply of propellants (nitrogen tetroxide/MMH) in pods on either side of tail fin. Additional propellant tanks can be fitted in aft end of cargo bay for manoeuvring up to 600 miles (960 km) altitude. Attitude control system concentrated in forward module in nose and aft system in each of OMS pods. Nose module contains 14 fixed jets of 870 lb (3.87 kN) thrust and two vernier jets of 25 lb (11.2 N) thrust arranged in three clusters and covered by hinged panels during launch and re-entry. Each aft system has 12 fixed jets and two vernier jets. Propellants same as OMS. Crew quarters comprises flight deck with positions for three crew, commander, co-pilot, mission specialist, plus passenger (scientist/payload specialist). Cabin below has galley, hygiene and sleeping facilities plus positions for three passengers. Also has airlock to payload bay. Main part of fuselage occupied by payload bay measuring 60 ft (18.3 m) long and 15 ft (4.6 m) wide with two-segment doors and TV-equipped manipulator arm. Design life is 100 missions or ten years
Payload: Design objectives include maximum of 65,000 lb (29,484 kg) into 115 mile (185 km) due-East orbit from Kennedy Space Centre; 25 000 lb (11 340 kg) into 312 mile (500 km) orbit at 55° from KSC; 32 000 lb (14 515 kg) into 115 mile (185 km) polar orbit from Western Test Range. Maximum return payload is 32 000 lb (14 515 kg)
Mission: Basic mission endurance defined as 49 man-days or seven days with standard three-man crew and four passengers. Can be extended to 30 days by substituting supplies for payload. Optimum turnaround time is 160 h including 2 h launch readiness
Programme: First re-usable satellite launcher/space transporter is scheduled to enter service in 1980. Initial plans call for five Shuttles with two operating from WTR and three from KSC. Prime contractor: Rockwell
Launch vehicle: Space Shuttle main engines-solid rocket boosters

Role: Lunar landing
Data: Luna 9
Height: approx 9 ft (2.7 m)
Diameter: approx 3 ft (0.9 m). Capsule approx 2 ft (0.6 m)
Launch weight: 3 491 lb (1 583 kg). Capsule approx 220 lb (100 kg)
Spacecraft: Carrier vehicle for soft landing capsule comprises conical propulsion module and cylindrical instrument compartment. In-flight control equipment located in jettisonable boxes attached to instrument compartment on top of which is fitted landing capsule wrapped in thermal blanket.
Mission: Luna 9, launched on January 31, 1966, made first ever survivable touchdown on another world, landing in Ocean of Storms on February 3. Direct approach initiated 1 hr before touchdown at 5 146 miles (8 280 km) altitude with craft assuming vertical attitude in relation to Moon. At 48 seconds and 47 miles (76 km) range, two external boxes jettisoned to reduce weight and main engine ignited for braking burn. Terminal velocity was 18-20 fps (5.5-6.0 m/sec). Capsule ejected just before touchdown to make free-fall landing. Upper hemisphere opened outward in four petals to stabilise craft. Four antennae erected on rim and central rotating mirror raised to reflect panoramic views to internal TV camera. Batteries provided 9 hr of communications over four days, altogether producing single 360° panorama of area
Programme: Luna 9 was first successful soft lander in varied series for lunar exploration spacecraft totalling at least 29 launches between January 1959 and end of 1974, of which 14 were wholly or partially successful. Series divides into three generations: 1) Luna 1 impact probe missed Moon but Luna 2 impact and Luna 3 photographic flyby were successful; 2) Soft landing and orbiter series all used same spacecraft bus. Poor start included failures by Lunas 4, 5, 6, 7 and 8 plus launch failures Cosmos 60 and 111 and one unannounced. Luna 13 was improved Luna 9 repeat and Lunas 10, 11, 12 and 14 all carried orbital survey packages instead of landing capsules; 3) robot explorer series included sample collector, Lunokhod rover plus advanced orbiters. Launch failures were Cosmos 300 and 305. Landing failutes were Luna 15, 18 (sample collector) and 23 (improved sample collector). Successes were 16 and 20 (sample collectors), 17 and 21 (Lunokhods) and 19 and 22 (orbiters). Programme appears to be continuing
Launch vehicle: A-2-e

LUNA SAMPLE COLLECTOR

Role: Lunar landing
Data: Luna 20
Height: Approx 12 ft 6 in (3.8 m)
Diameter: Approx 13 ft (4.0 m) across landing pads
Spacecraft: Dual spacecraft system consists of standard descent stage surmounted by ascent stage and re-entry capsule.
Descent stage: Serves as translunar mother ship and lunar landing craft also used for Lunokhod automatic rovers. Four spherical propellent tanks are linked by cylindrical structural skirting and enclose non-throttleable liquid-fuel main engine. Engine nozzle is flanked by two small solid-fuel terminal braking motors. Assembly surmounted by annulus with protruberances on three sides containing descent stage electronics. Fourth side has remotely controlled arm with drillhead containing 14 in (35 cm)-long sample drill. Craft is three-axis stabilised and has four large vernier thrusters underneath to control descent. Power provided by chemical batteries
Ascent stage: Simple carrier vehicle for return flight comprises cylindrical instrument compartment with three propellent tanks fitted underneath. Liquid-propellent engine attached to central tank is fitted within annual cavity of descent stage electronics compartment. Two large vernier thrusters fitted to outside tanks and two on outriggers provide attitude control in two axes and smaller role thrusters also carried on outriggers. Craft also powered by chemical batteries
Re-entry capsule: Spherical container about 20 in (50 cm) in diameter is completely covered in ablative material for direct approach re-entry. Side hatch allows insertion of drill core and top hatch is jettisoned to release parachutes and homing beacon antennae. Two gas-inflated cylindrical balloons project through top hatch either to ensure parachute deployment or to stabilise capsule in case of water landing
Mission: Luna 20 launched in February 1972, touched down on edge of Sea of Fertility on February 21 after three days in lunar orbit during which perilune was lowered to 13 miles (21 km) for orbital science programme. Descent initiated by 267 sec burn and then free fall to 2,500 ft (760 m) altitude. Second burn then continued to 66 ft (20 m) altitude before cutout and ignition of terminal braking rockets which burnt out at 7 ft (2.0 m) altitude. Descent time was 6 min and touchdown speed was 8 ft (2.5 m)/second (figures quoted for Luna 16). Surface stay of 26.5 h included surface science programme. Sample weight unannounced but Luna 16 sample weighed 3½ oz (10 gm)
Programme: Second successful use of robot sample collecting system first demonstrated by Luna 16 in September 1970. Another attempt, Luna 18, crash landed in September 1971. Luna 15 (July 1969) may have been first attempt also ending in crash landing. Latest attempt was Luna 23 (October 1974) which was unusable after bad lunar landing
Launch vehicle: D-1-e

LUNOKHOD

Role: Lunar surface exploration
Data: Lunokhod 2
Height: About 4 ft 4 in (1.35 m). Wheelbase: 7 ft 3 in (2.2 m)
Diameter: 7 ft (2.15 m). Track: 5 ft 3 in (1.6 m)
Weight: 1 840 lb (840 kg)
Spacecraft: Remotely controlled lunar roving vehicle has circular tub-shaped pressurised instrument compartment as main body. Convex top has fitted thermal insulation cover which hinges open rearward up to 180° during lunar day to expose solar array covering under surface. Vehicle rides on eight independently powered mesh wheels each 20 in (51 cm) in diameter and trails odometer wheel at rear. A spiral omnidirectional low-gain antenna and a helical directional medium-gain antenna are erected above the front. Elevated panoramic TV camera provides general viewing and pair of low-level stereoscopic cameras provide driving vision for five-man operating crew. Vehicle can travel at constant slow speed or in short individually commanded steps and can also reverse. Carrier vehicle is standard descent stage (see Luna sample collector) but modified by relocation of instrument compartments so that carriage platform can be fitted. Ramps lower at either end in case of impeded exit in one direction

Payload: Main experiment is "Rifma" spectrometer carried on rod between front wheels to make en route X-ray analysis of soil. Probe device tests physical-mechanical properties of soil. Also carried are proton, electron and alpha particle detectors, an X-ray telescope and a French-built laser reflector
Mission: Launched on January 8, 1973, Luna 21 touched down in Sea of Serenity on January 16 after four days in lunar orbit. Lunokhod 2 survived four lunar "nights" and covered 23 miles (37 km) before final shutdown at beginning of July 1973
Programme: Second Lunokhod similar to Lunokhod 1/Luna 17 which operated in Mare Imbrium from November 1970 to October 1971 and covered six miles (9.6 km) before expiry of isotope heater allowed internal systems to freeze up. No reason announced for shorter life of Lunokhod 2 but both exceeded planned operational life of three months. Modifications on Lunokhod 2, including improved camera system, added 173 lb (79 kg). Luna 15, Cosmos 300 and Cosmos 305 (July, September and October 1969) may also have failures either of Lunokhods or Lunar sample collectors
Launch vehicle: D-1-e

Role: Interplanetary engineering (manned precurser)
Data: Zond 8
Height: About 17 ft 2 in (5.3 m). Re-entry module 7 ft 2 in (2.2 m)
Diameter: 7 ft 3 in (2.35 m)
Launch weight: About 12 000 lb (5 443 kg)
Spacecraft: Modified Soyuz craft comprising cylindrical service module and dome-shaped re-entry module but without orbital module (see Soyuz). Power supplied by pair of solar arrays deployed from sides of service module. Propulsion system probably similar to Soyuz with main engine developing 880 lb (3.9 kN) thrust, plus back-up engine. Re-entry module adapted to carry scientific payload and was fitted with high-gain dish antenna
Payload: Re-entry module capable of supporting human crew but fitted with colour and black and white TV cameras plus laser projector
Mission: Launched on October 20, 1970, Zond 8 rounded the Moon on October 24 at distance of 696 miles (1 120 km) and splashed down in Indian Ocean 450 miles (724 km) south of Chagos Archipelago on October 27. First Zond to make direct ballistic re-entry over North Pole. Also made first en route pictures of Earth during outbound flight and was first Soviet mission to have objectives announced at beginning of flight
Programme: Total of eight Zond-designated flights launched between April 1964 and October 1970 consisted of two groups. First three were engineering flights for unmanned interplanetary flights and probably resembled early Mars type bus. Zond 1 was Venus flyby mission. Zond 2 (November 1964) was Mars flyby mission. Both had communications failure en route. Zond 3 (July 1965) sent to Mars orbit, photographing back of Moon and made repeated long-distance test transmission of pictures. Zond 4-8 (March 1968) and subsequent missions were all Soyuz-type manned precursers. Zond 4 fired at simulated Moon and entered solar orbit. Zond 5, 6 and 7 made low-g skip entries over South Pole after circumlunar flights. Zond 5 and 8 recovered in Indian Ocean. All capsules carried engineering, space science or biological payloads. Soviet statements strongly suggest application to plans (now discontinued) for manned lunar missions. Programme apparently concluded
Launch vehicle: D-1-e Proton

Role: Planetary orbiter/lander
Data: Mars 3
Height: Approx 13 ft 6 in (4.1 m) with lander assembly
Diameter: Aprox 6 ft 6 in (2.0 m) across base. Array span: approx 19 ft 6 in (5.9 m)
Launch weight: 10 251 lb (4 650 kg)
Spacecraft: Dual spacecraft comprises orbiter/mother ship plus lander assembly. Orbiter is built round cylindrical centre section containing hypergolic-propellant tanks for braking and manoeuvring engine attached to base of cylinder. Engine is completely enclosed within large toroidal compartment containing spacecraft and experiment electronics and linked to upper section by enshrouding skirt. Top part of cylinder detaches as part of lander assembly. Large radiator panel projects from one side. Two deployable solar arrays each have lander communications antennae on the back, plus attitude control jets on the tips. One panel also carries magnetometer boom and antenna for Stereo experiment. Communications use high-gain dish antenna of about 100 in (2.5 m) diameter and pair of low-gain omnidirectional spiral antennae
Payload: Lander assembly (detailed separately) plus orbital science package comprising two TV cameras, infra-red radiometer, infra-red photometer, ultra-violet (water vapour) spectrometer, visible light photometer (for studying albedo and atmosphere) and a radio-brightness sensor for surface temperature measurement
Mission: Mars 2 entered 18 h Mars orbit in November 1971 and Mars 3 entered 11-day orbit in December. Began similar mission to Mariner 9, also in Mars orbit. Surface mapping programme delayed by dust storm until February 1972 but both probes continued operating until September, including two-week shut-down during solar occultation period. Some data exchanged with USA during mission
Programme: Total of 12 Mars attempts launched in five windows between October 1960 and August 1973 includes five launch failures and limited successes. Mars 1 (November 1962) suffered communications failure mid-flight. Mars 2 and 3 launched in May 1971 with third craft (Cosmos 419) lost. Both achieved Mars orbit but Mars 2 capsule made "hard landing" and Mars 3 capsule ceased signalling after 20 seconds of surface TV. Mars 4, 5, 6 and 7 launched in 1973 window, first pair being orbiting relay craft and second pair carrying new large landers for release during flyby. Mars 4 failed to enter orbit, but Mars 5 achieved scheduled orbital photography mission. Mars 6 released lander which ceased communicating just before touchdown and Mars 7 lander missed planet. No spacecraft details released of last four
Launch vehicle: D-1-e

MARS LANDER

Role: Planetary lander
Data: Mars 3 lander
Diameter: Capsule approx 4 ft (1.2 m). Heatshield approx 9 ft 6 in (2.9 m)
Launch weight: Complete assembly estimated at 2 000 lb (907 kg)
Spacecraft: Lander is near-spherical capsule probably similar in principle to Luna 9 capsule. Wide, conical heat shield provides aerodynamic braking. Capsule sits in large annular parachute compartment below which extends a cylindrical orbiter-attachment section. Cylinder encloses solid-fuel spacecraft separation engine and descent control instruments and holds base of heat shield support struts. Capsule also enclosed by shielding linking annulus to rim of heatshield
Payload: Mass spectrometers for chemical analysis of atmosphere during descent as well as instruments for determining chemical content and physical and mechanical properties of surface soil. Also carries TV camera, wind anemometer, thermometer and pressure gauge. Data transmitted to orbiter for relay to Earth
Mission: Mars 3 lander assembly released just before orbit insertion of Mother craft. Penetrated atmosphere at about 13 500 mph (21 600 kph). Drogue parachute explosively deployed to extract main parachute while still at supersonic velocity after jettison of heat shield. Parachute and other components jettisoned at 70-100 ft (20-30 m) altitude and final braking achieved by small solid-fuel charge probably within capsule itself. TV camera activated 90 sec after touchdown and transmitted indecipherable pictures for 20 sec before capsule suffered complete communications failure
Programme: Mars 2 and 3 believed to be identical orbiter/lander combinations. Mars 2 capsule made "hard landing", presumed to be soft-landing failure, on November 27, 1971 making first ever contact with Mars surface. First known attempt at landing. Mars 3 made apparently nominal descent and landing on December 2. Mars 6 and 7 carried new, larger landers of which no details have been released. Mars 6 lander made nominal descent but ceased communicating just before touchdown. Mars 7 lander released prematurely, missed planet and entered solar orbit (see Mars)
Launch vehicle: Mars 3 orbiter/D-1-e

VENERA

Role: Planetary landing
Data: Venera 8
Height: Approx 10.5 ft (3.2 m) overall. Capsule approx 3 ft 3 in (1.0 m)
Diameter: Approx 3 ft (0.9 m). Array span approx 11 ft (3.35 m)
Launch weight: 2 596 lb (1 177 kg). Capsule 1 089 lb (494 kg)
Spacecraft: Modular spacecraft follows pattern of Luna 9-type craft in having cylindrical, sealed instrument compartment with propulsion module attached at one end and landing capsule attached to the other. Deployable wire-mesh high-gain dish antenna and omnidirectional low-gain spiral antenna fitted to "front" and pair of solar panels extend from "sides" and anti-reflection panel carrying navigation sensors attached to "back". Low-gain antenna serves all communications for first half of flight after which dish antenna is orientated for command and data communications
Payload: Venus landing capsule is roughly spherical with weighted base for self-righting and jettisonable top which releases parachute and exposes antennae and sensors attached outside sealed instrument compartment. Pressure and temperature protection reduced from that of Venus 7 capsule and instrumentation increased in continuing search for optimum balance. Thermal protection included use of heat sink and cryogenic fluids to "freeze" capsule just before atmospheric entry. Equipped for atmospheric analysis, wind, temperature and pressure measurement during descent and on surface plus X-ray spectrometer for chemical soil analysis. Carrier vehicle also equipped with sensors for measurement of solar radiation en route
Mission: Launched on March 2, 1972, Venera 8 approached Venus on July 22 and released capsule automatically on contact with outer atmosphere. Aerodynamic braking reduced speed from 26 000 mph (41 800 km/h) to 560 mph (900 km/h) at record low altitude of 40 miles (64 km) before parachute released to minimise exposure time to heat. Made first landing on Sun side of planet where it transmitted data for 50 min before overheating. Ejected flat spiral antenna after touchdown in case capsule should sink below surface. Carrier vehicle destroyed in atmosphere
Programme: Venus 8 was first of two sister craft launched in 1972 window. Total of 21 Venus probe attempts made by USSR between February 1961 and end of 1975. Ten detected as failing to leave Earth orbit and were unannounced or given Sputnik or Cosmos designations. Chronology by launch windows — 1961: Sputnik 4 and Venera 1 (signals lost en route); 1962: three unannounced; 1963: Cosmos 21 (not in window — possibly engineering test in Earth orbit); 1964: Zond 1 (signals lost en route); 1965: Venera 2 (signals failure), Venera 3 (successful impact shot), Cosmos 96; 1967: Venera 4 (capsule signals lost at 15 miles altitude), Cosmos 167; 1969: Venera 5 and 6 (capsule signals lost during descent); 1970: Venera 7 (capsule transmitted for 23 min on surface), Cosmos 359; 1972: Venera 8, Cosmos 482; 1975: Venera 9 and 10 (obtained first surface photograph)
Launch vehicle: A-2-e

SOLAR PIONEER

Role: Solar orbital survey
Data: Pioneer 9
Height: Main body 2 ft 11 in (0.9 m)
Diameter: 3 ft 1 in (0.95 m)
Launch weight: 148 lb (67.1 kg)
Spacecraft: Cylindrical main body has narrow band around middle to expose sensors and experiment apertures with remaining surface covered in solar cells generating 60 W. A 52 in (1.3 m)-long telemetry and command antenna extends from top and three 64 in (1.6 m)-long side booms carry an experiment, a wobble damper and attitude-control thrusters with nitrogen gas supply. Spacecraft is spin-stabilised at 60 rpm
Payload: Eight experiments in 39.5 lb (17.9 kg) payload include magnetometer on boom plus sensors to measure solar wind, cosmic dust, cosmic ray particles, electric fields and electron density. Also equipped to track data from all other Pioneers to refine measurements of celestial mechanics
Mission: Pioneer 9 launched in November 1968 (in dual launch with TETR 2, jettisoned into Earth orbit). Solar orbit of 0.75 AU to 1.0 AU has lowest point at 70 million miles (113 million km) and closest to Sun of any in Pioneer series. Continued operating until 1974
Programme: Series of five identical spacecraft with similar payloads launched between December 1965 and August 1969 to collect data on "quiet Sun" years. Pioneers 6, 7, 8 and 9 successfully placed in solar orbit. Last in series, Pioneer E, lost in launch failure. Spacecraft all far exceeded design lives and provided bonus of solar "weather forecasting", giving up to 15 days' warning of solar activity that might endanger Apollo crewmen. Prime contractor: TRW
Launch vehicle: TA Improved Delta

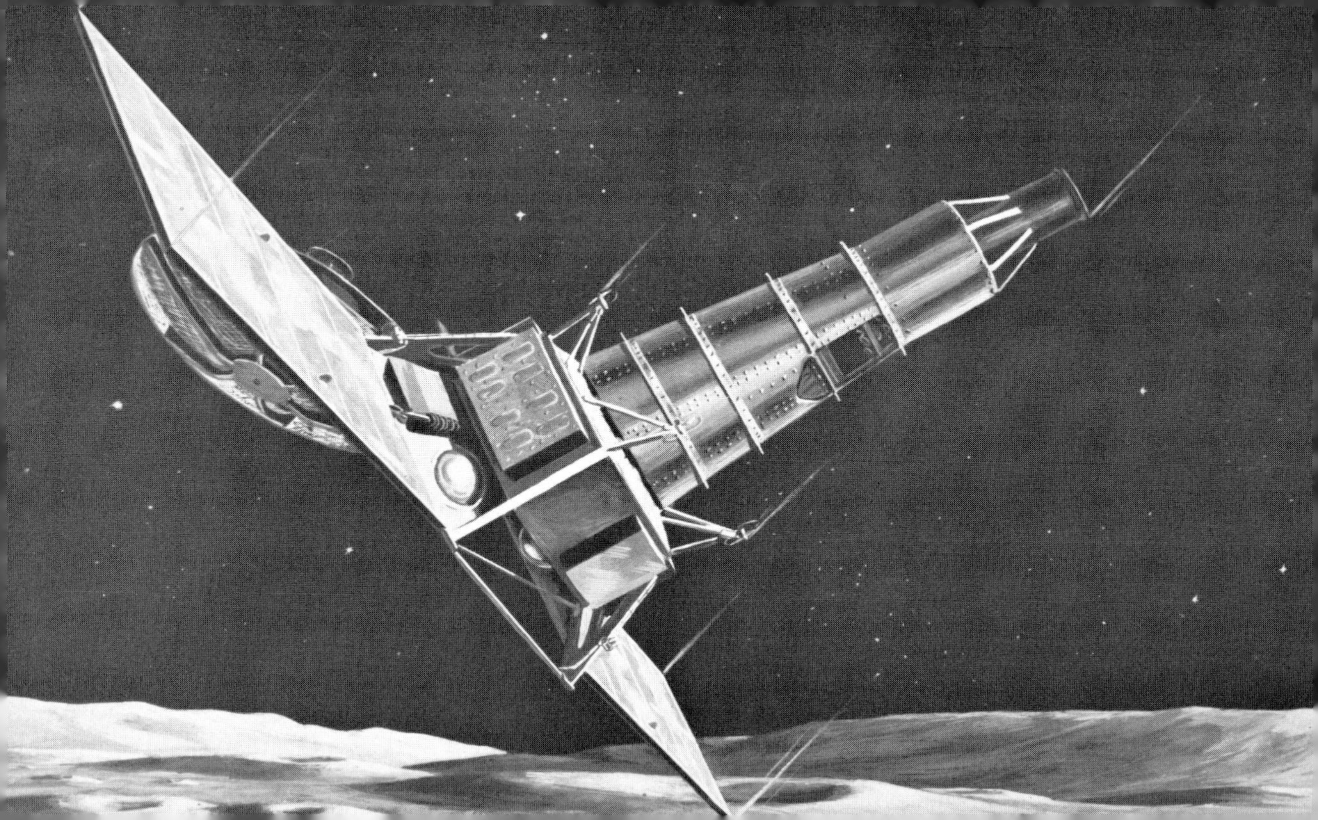

RANGER

Role: Lunar impact
Data: Ranger 9
Height: 10 ft 3 in (3.1 m)
Diameter: 5 ft (1.5 m). Array span: 15 ft (4.5 m)
Launch weight: 809 lb (367 kg)
Spacecraft: Hexagonal main structure has electronics boxes attached to each of six sides and encloses 50 lb (222 N)-thrust course-correction engine. High-gain dish antenna and two solar arrays generating 200 W are hinged to base of hexagon on which stands a 5 ft (1.5 m)-high slightly conical camera shroud surmounted by cylindrical omnidirectional low-gain antenna
Payload: Self-contained TV subsystem includes its own batteries, radios and shroud in 380 lb (172 kg) package. Six cameras view through 13 in (33 cm)-high aperture in side of shroud and comprise two wide-angle and four narrow-angle cameras with fixed focus but able to operate between distances of 1 100 miles (1 770 km) and ½ mile (0.8 km) from surface
Mission: Launched in March 1965, Ranger 9 made shallow approach to impact in Alphonsus crater, returning 5 814 photographs during final 1 500 miles (2 413 km). Final pictures at 2 000 ft (610 m) showed resolution of 10 in (25 cm). Impacted 2.76 miles (4.4 km) from target. Unique feature was live TV broadcast of 200 pictures during final descent phase
Programme: Ranger was first US lunar science exercise following fruitless Pioneer lunar impact programme. Nine spacecraft launched between August 1961 and March 1965. First two were engineering flights not aimed at Moon but both failed to leave Earth orbit. Next three carried three cameras as well as 25 in (63 cm) diameter instrumented balsa sphere on top intended for survivable landing. Rangers 3 and 5 missed Moon and 4 crashed on far side. Last four similar to Ranger 9 were all successful except for camera failure on Ranger 6. Prime contractor: Jet Propulsion Laboratory
Launch vehicle: Atlas-Agena B

LUNAR ORBITER

Role: Lunar orbital survey
Data: Lunar Orbiter 5
Height: 5 ft 6 in (1.7 m)
Diameter: 5 ft (1.5 m). Array span: 12 ft 2 in (3.7 m)
Launch weight: 860 lb (390 kg)
Spacecraft: Circular structural "deck" carries truncated cone containing instruments and egg-shaped pressure shell for camera system. On top are hypergolic-propellant tanks and 100 lb (445 N)-thrust main engine with nozzle projecting through heatshield. Entire assembly enclosed within thermal blanket. Four square solar panels, a high-gain dish antenna and low-gain mast antenna all hinge to horizontal position around deck
Payload: 150 lb (69 kg) photographic laboratory contains one wide-angle and one telephoto camera viewing through quartz window plus film for at least 212 dual exposure frames and processing equipment. Also carries proton radiation and meteoroid detectors
Mission: Orbiter 5 was placed in near-polar Moon orbit at 125 x 3,700 miles (201 x 5,953 km) to complete lunar far side photography and then lowered to 60 x 930 miles (96 x 1,496 km) for Apollo site photography. Best resolution was 6 ft (1.8 m). Completed main mission in one month and then used by Nasa as tracking target in third slightly higher orbit
Programme: Highly successful programme of five identical spacecraft launched at three-month intervals between August 1966 and August 1967 to photograph potential Surveyor and Apollo landing sites and to assess risk from meteoroids and solar radiation in lunar orbit. Main role completed by first three Orbiters from near-equatorial orbit and last two were placed in near-polar orbits with predominantly scientific tasks. All spacecraft intentionally crashed into Moon at end of mission to prevent risk of accidental reactivation producing radio interference. Programme achieved photography of 99.5 per cent of lunar surface and produced first detailed maps of far side as well as first US spacecraft to orbit Moon and first photographs of Earth from the Moon. Prime contractor: Boeing
Launch vehicle: Atlas-Agena D

SURVEYOR

Role: Lunar landing
Data: Surveyor 7
Height: 10 ft (3.0 m)
Diameter: 14 ft (4.3 m) across landing pads
Launch weight: 2 288 lb (1 038 kg). Landed weight 625 lb (283 kg)
Spacecraft: Triangular frame provides mounting points for three deployable landing legs, engines, tanks and electronics compartments. Central mast carries square high-gain planar-array antenna and square solar panel generating 89 W. Propulsion system comprises three vernier engines burning hypergolic propellants and throttleable between 30-104 lb (133-463 N) thrust for course corrections and terminal braking. Main braking provided by 10 000 lb (44.5 kN)-thrust solid-fuel motor weighing 1 444 lb (655 kg) which slows spacecraft from 6 000 mph (9 654 km/h) to 250 mph (402 km/h) at 37 000 ft (11 278 m) and then jettisoned. Touchdown speed is 12 ft (3.6 m)/sec
Payload: Vertically mounted TV camera points at rotating and swivelling mirror. Surface scoop can be extended up to 5 ft (1.5 m) and pivoted 112°. Scoop tests physical and mechanical properties of soil, positions samples for TV viewing and subjects them to chemical analysis by alpha-particle scattering device. Scoop can also reposition alpha-scattering instrument. Also equipped for several technology experiments
Mission: Surveyor 7 landed on rim of Tycho crater on January 7, 1968, closed down for lunar night on January 23 and reactivated on February 6. Signals lost on February 20 but objectives of mainly scientific mission already achieved. Returned 21 000 high-resolution pictures
Programme: Seven-spacecraft programme was designed to master soft-landing techniques and to make close study of possible Apollo landing sites. Basic plan was for four engineering flights followed by three operational flights. Programme preceded by undesignated proof-of-concept launch of test model at Moon's orbit. Surveyor 2 lost in crash landing and Surveyor 4 had signal failure during final descent but may have landed automatically. Scoops carried on Surveyors 3, 4, 6 and 7. Surveyor 6 also made first ever take-off from Moon when vernier engines were reignited to lift it 10 ft (3.0 m) from surface and land it again 18 ft (2.4 m) away. Prime contractor: Hughes
Launch vehicle: Atlas-Centaur

MARINER VENUS

Role: Planetary fly-by
Data: Mariner 5
Height: (Main body) 1 ft 8 in (0.5 m). Overall 9 ft 6 in (2.9 m)
Diameter: 4 ft 2 in (1.25 m). Array span: 18 ft (5.5 m)
Launch weight: 540 lb (245 kg)
Spacecraft: Modified Mariner 4 backup spacecraft has standard Mariner octagonal main body with electronics boxes fitted inside each of seven sides enclosing hydrazine tank and 50 lb (222 N)-thrust course-correction engine. Engine nozzle protrudes through eighth side. Main body surmounted by elliptical high-gain antenna and 7 ft 4 in (2.2 m)-long low-gain omnidirectional mast antenna. Four solar panels hinged to top edge of body generate 370 W at launch and 555 W at Venus. Three-axis stabilisation maintained by nitrogen-gas thrusters at ends of solar panels. Spacecraft has deployable sunshield fitted to underside
Payload: Two groups of experiments. En route data provided by magnetometer (on low-gain antenna mast), trapped radiation detector and radio propagation. Planetary data from ultra-violet photometer and occultation of S-band telemetry signals
Mission: Mariner 5 launched on June 14, 1967 and passed Venus at 2,480 miles (3,990 km) distance on October 19. Arrived one day after Venera 4 capsule entered atmosphere. Mariner 5 data, which provided most accurate atmospheric temperature and pressure figures of the time, also confirmed that Soviet capsule had expired during descent and that final figures were at 15.5 miles altitude
Programme: Mariner 5 was second successful US Venus probe attempt and is still closest approach to planet by any US spacecraft. Preceding missions were Mariners 1 and 2 which were modified Rangers from which later Mariner evolved. Both launched in 1962 window but Mariner 1 lost during launch. Mariner 2 achieved flyby at 21,600 miles (34,754 km) range. Only subsequent attempt is Mariner 10 which provided data plus first TV pictures of planet during swingby manoeuvre at 3,573 miles (5,749 km) range en route for Mercury (see Mariner Mercury). Prime contractor: Jet Propulsion Laboratory
Launch vehicle: Atlas-Agena D

MARINER MERCURY

Role: Planetary flyby
Data: Mariner 10
Height: Main body 1 ft 6 in (0.45 m)
Diameter: 4 ft 6 in (1.39 m)
Launch weight: 1 108 lb (503 kg) plus jettisonable adaptor
Spacecraft: Standard Mariner main body is 40 lb (18 kg) octagonal structure with electronics compartments fitted inside each of eight sides. Encloses a spherical tank containing hydrazine monopropellent for 50 lb (222 N)-thrust course-correction engine fitted on top of tank. Engine nozzle projects through centre of deployable sunshield. Other projections include 20 ft (6.1 m)-long magnetometer boom, 9.5 ft (2.9 m) long low-gain antenna mast and 54 in (1.37 m)-diameter high-gain dish antenna on outrigger assembly. Two 9 ft (2.7 m)-long solar panels can be angled 76° away from Sun in five steps to maintain constant surface temperature during approach and generating 475 W at Mercury. Spacecraft is three-axis stabilised with duplicate nitrogen-gas attitude-control system using thrusters on array tips and on dish antenna and magnetometer boom outriggers
Payload: 172 lb (78 kg) payload includes instrumentation for six experiments arrayed around spacecraft. Twin TV cameras and ultra-violet airglow spectrometer are mounted on scan platform controllable in two axes and fitted to side opposite engine. U-V occultation spectrometer and charged particle telescope are fixed to main body. Two magnetometers are carried on single boom and plasma detectors are mounted on short, motor-driven boom. A seventh experiment involves use of signals from spacecraft's S-band and X-band transmitters
Mission: Mariner 10 launched in November 1973. Made first-ever use of planetary swing-by technique using gravity of Venus to place it on correct trajectory for Mercury flyby at 463 miles (945 km) altitude in late March 1974. Took first close-up pictures of Venus en route. Entered solar orbit which brought it back for second Mercury encounter in September 1974 and third in March 1975 before fuel exhaustion ended useful life
Programme: Mariner programme uses near-standard spacecraft for flyby and orbital study of inner planets. Ten Mariners launched since July 1962 include four to Venus — Mariners 1 (launch failure), 2, 5 and 10. Remainder went to Mars and also included one launch failure (Mariner 8). Prime contractor: (Mariner 10) Boeing
Launch vehicle: Atlas-Centaur

MARINER MARS

Role: Planetary orbital survey
Data: Mariner 9
Height: Main body 1 ft 6 in (0.45 m). Overall 7 ft 6 in (2.3 m)
Diameter: 4 ft 6 in (1.4 m). Array span 22 ft 7 in (6.9 m)
Launch weight: 2 200 lb (1 000 kg). Empty weight 1 200 lb (544 kg)
Spacecraft: Three-axis stabilised spacecraft has octagonal basic structure weighing 40 lb (18 kg) with four solar panels attached to upper rim generating 800 W at launch, 500 W at Mars. Duplicated attitude-control jets on all panel tips. All electronics in boxes fitted to inside of each of eight structural facets. Two spherical tanks containing monomethyl hydrazine and nitrogen tetroxide hypergolic propellents attached to top of structure and surmounted by 300 lb (1.33 kN)-thrust manoeuvring and orbit-insertion engine. Computer controls basic mission without ground command except for manoeuvres. Communications use 40 in (1.02 m)-diameter high-gain dish antenna, 57 in (1.45 m)-long low-gain mast antenna and medium gain horn antenna
Payload: Four experiments mounted on a scan platform fitted to base of main structure and which can be rotated about two axes. Two TV cameras — one wide angle, low resolution (3 280 ft, 1 000 m) and one narrow angle. high resolution (328 ft, 100 m) — ultra-violet spectrometer, infra-red radiometer and infra-red interferometer spectrometer
Mission: Mariner 9 reprogrammed after Mariner 8 loss to perform compromise dual mission. Launched on May 30, 1971 and entered 12 h highly eccentric Mars orbit on November 13. Dust storm obscured surface for first two months but cleared in time for most of mission to be completed within 90 day programme. Spacecraft survived two-week shutdown during solar occultation period and continued operating until October 1972, by which time it had photographed 90 per cent of surface, comfortably exceeding 70 per cent coverage target. Also photographed Mars moons, Phobos and Deimos
Programme: Sixth US Mars probe launched since November 1964 (Mariners 3, 4, 6, 7, 8 and 9) and fourth success (Mariners 3 and 8 were launch failures). First success was Mariner 4 which provided first pictures of another planet (22 pictures) plus data during 6 118 mile (9 844 km)-distant flyby. Mariners 6 and 7 reduced flyby distance and increased picture supply. Mariners 8 and 9 were twin spacecraft programmed for complementary surface mapping missions (general and specific) and were first planetary orbiters. Prime contractor: Nasa Jet Propulsion Laboratory
Launch vehicle: Atlas-Centaur

VIKING

Role: Planetary orbital survey/landing
Data: Viking orbiter
Height: Launch configuration with lander 16 ft (4.9 m)
Diameter: Main body diagonally 7 ft 1 in x 8 ft 3 in (2.2 m x 2.5 m). Array span 33 ft (10 m)
Launch weight: Orbiter 5 125 lb (2 325 kg). Complete assembly 7 624 lb (3 458 kg)
Empty weight: Orbiter 2 028 lb (920 kg). Complete assembly 4 220 lb (1 914 kg)
Spacecraft: Based on Mariner design with main body as octagonal frame in shape of near-square, hollow diamond with truncated points and made up of 16 equipment bays. Four deployable solar arrays attached to short sides generate 620 W at Mars and carry attitude control jets on tips. Appendages to main body include 58 in (1.5 m) diameter X-band/S-band dish antenna, S-band low-gain mast antenna, navigation sensors and science scan platform (stowed during flight). Lander relay antenna attached to one solar panel. ACS nitrogen tanks occupy two equipment bays. Propellant pressuring tank enclosed within main body and surmounted by propulsion system comprising two 55 in (1.4 m)-long propellant tanks (nitrogen tetroxide/hydrazine) and 300 lb (1 334 N)-thrust manoeuvring and orbit-insertion engine. Opposite side has mountings and power connections for lander assembly
Payload: Lander assembly plus 152 lb (69 kg) orbital science payload consisting of a pair of high resolution TV cameras, an infra-red (atmospheric-water) spectrometer and an infra-red (thermal mapping) radiometer mounted on steerable scan platform
Mission: Nominal journey time is 305 days, extending as launch window progresses to 360 days. Terminal engine burn of nearly one hour required from insertion into 940 x 20 400-mile (1 500 x 32 600 km) orbit with periapsis lasting 24.6 h. Period coincides with Mars "day" and is therefore synchronous. At least ten days allocated to site selection before lander is released. Orbiter then manoeuvres periapsis to coincide with landing site
Programme: First US landing attempt on another planet comprises two identical spacecraft launched in August and September 1975, are scheduled to reach Mars several weeks part. Objective is to land first craft in time to provide accurate advance data before second craft enters orbit
Launch vehicle: Titan IIIE-Centaur

VIKING LANDER

Role: Planetary landing
Data: Viking lander
Height: Main body 18 in (0.46 m). Capsule 6 ft 4 in (1.9 m)
Diameter: Main body (width) 4 ft 11 in (1.5 m). Capsule 12 ft (3.7 m)
Launch weight: (with capsule) 2 771 lb (1 257 kg)
Empty weight: 2 464 lb (1 118 kg)
Landed weight: 1 269 lb (576 kg)
Spacecraft: Mars landing vehicle has hexagonal main body in shape of triangle with truncated corners. Contains electronics and instruments. Three landing legs each 4 ft 3 in (1.3 m) long attached to short sides with engine-shutdown switch on each footpad. Three terminal descent engines and two hydrazine tanks attached to long sides. Engines are throttleable to 585 lb (2.6 kN) thrust each through 18-nozzle clusters to minimise surface disturbance. Radar altimeter antenna projects from one corner and roll thrusters are attached to main propellant tanks. Two Snap-19-type radioisotope thermoelectric generators under thermal screens behind tanks produce 57 W and provide instrument compartment heating. Other appendages include UHF antenna (for orbiter relay signals), S-band low-gain antenna and 30 in (0.76 m)-diameter high-gain S-band antenna for direct communications back-up facility. Furlable sample-collector boom, meteorology boom, instrument projections and apertures fitted to top of main body. During flight, lander is powered by orbiter and carried, with appendages retracted, inside descent capsule, itself enclosed in pressurised, hermetically sealed, glass-fibre "bioshield" to preserve sterility. Lower half of capsule is "aeroshell" which is insulated heatshield for aerodynamic braking. Aeroshell fitted with four thrusters on rim and two hydrazine tanks to provide de-orbit retrothrust and descent attitude control. Upper half contains morter-deployed 37 ft (11.3 m)-diameter parachute
Payload: Entry science uses four sensors in aeroshell heatshield for atmospheric analysis during decent. Similar sensors on lander continue study after aeroshell jettison. Surface science payload includes two cameras with rotating and swivelling mirrors, integrated biology instrument which is fed by scoop to provide three different soil analyses, a similar device for inorganic chemical analysis, molecular analyser, a boom carrying meteorological sensors, a seismometer, and magnets on sampling scoop
Mission: Atmospheric entry begins 2-5 h after separation at 152 miles (244 km) altitude. Aerodynamic braking to about 21,000 ft (6,400 m) and 993 mph (1,598 km/h) before aeroshell jettison and parachute release. Parachute jettison at 4,100 ft (1,250 m) and 253 mph (407 km/h). Terminal engines fire until touchdown at 5.5 mph (8.8 km/h). Descent time 5-10 min. Planned minimum operational period is 90 days
Programme: Two Viking landers scheduled to make first US planetary landings several weeks apart in July-August 1976 (see Viking orbiter)
Launch vehicle: Titan-Centaur/Viking orbiter

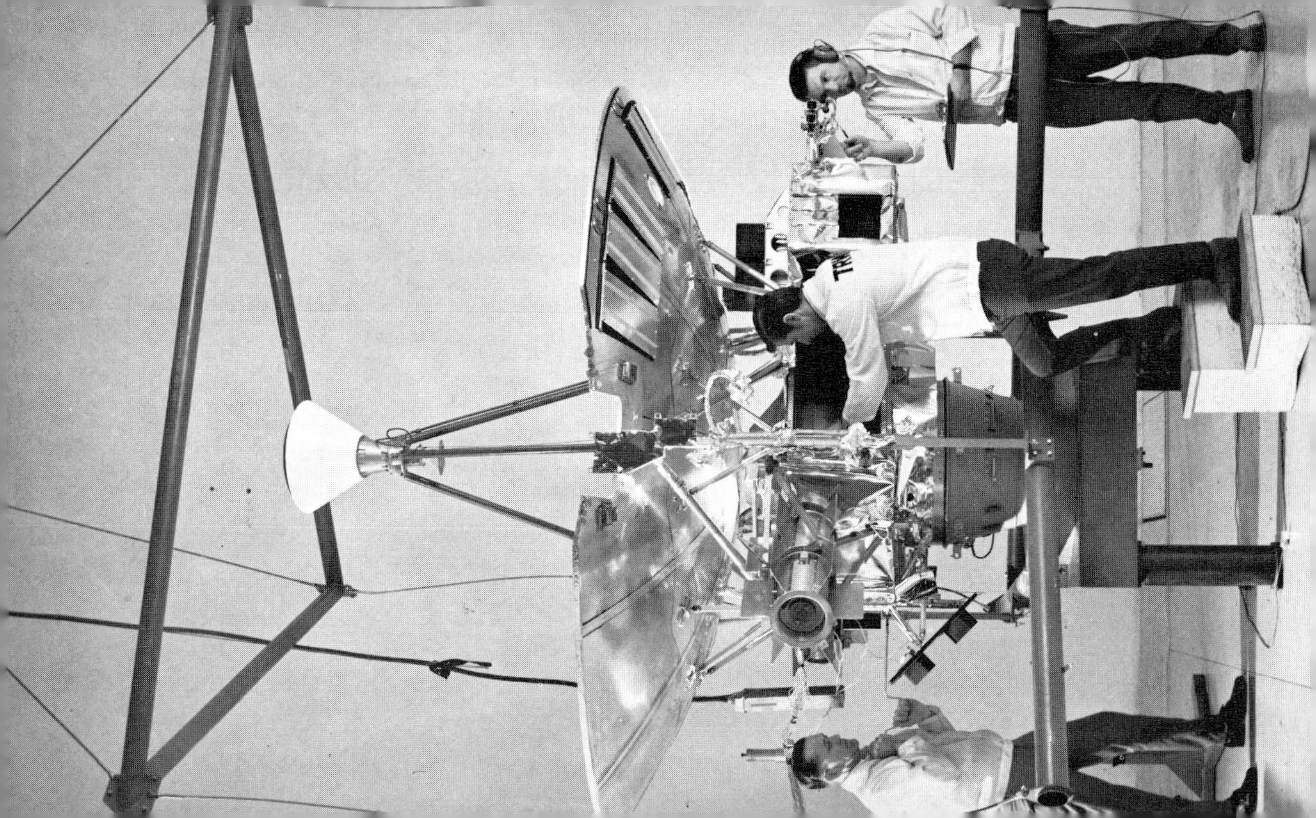

(USA-NASA) PIONEER JUPITER

Role: Planetary fly-by
Data: Pioneer 11
Height: 9 ft 6 in (2.9 m) overall
Diameter: 9 ft (2.7 m) across dish antenna
Launch weight: 570 lb (259 kg)
Spacecraft: Main spacecraft body and instrument compartment is regular hexagon measuring 28 in (71 cm) x 14 in (35 cm) per side. Attached to one side is irregular hexagon containing scientific payload. Upper face carries 9 ft (2.7 m)-diameter high-gain dish antenna with medium-gain horn antenna mounted at its focal point on three struts. Launcher mounting ring and low-gain mast antenna project from lower face. Two 9 ft (2.7 m)-long three-rod trusses each carry a SNAP 1a radioisotope thermoelectric generator (RTG), together generating 155 W (140 W at Jupiter). Third boom, 21 ft 6 in (6.6 m) long, carries magnetometers. Spacecraft is spin-stabilised at 4.8 rpm. Six hydrazine jets of 0.4-1.4 lb (1.8 N-6.2 N)-thrust each are mounted on dish-antenna rim to provide normal attitude control in three axes. Thrusters can also be fired symmetrically, roll thrusters controlling spin and fore- and aft-pointing jets providing velocity changes for course correction. Overall design provides lightness and simplicity
Payload: Main experiment compartment contains nine instruments of 66 lb (30 kg), 13-experiment payload. Included are cosmic-ray telescope, infra-red radiometer, charged particle detector, trapped radiation detector, ultra-violet photometer, geiger-tube telescope, imaging photopolarimeter, plasma analyzer, and (on boom) a flux-gate magnetometer (not on Pioneer 10) and a vector helium magnetometer. Asteroid-meteoroid detector is attached to main body opposite payload compartment and neteoroid sensor panels are fitted to underside of dish antenna. Two experiments use spacecraft's radio signals. Engraved plaque identifies place of origin in symbols and binary code in case of interception by intelligent extraterrestrials
Mission: Pioneer 11 launched on April 5, 1973, and reached Jupiter on December 3, 1974. Passed within 26 600 miles (42 800 km) of upper cloud layer during four-day flyby, or three times as close as Pioneer 10. Close approach allowed gravity of planet to redirect spacecraft to reach Saturn in October 1979
Programme: Two near-identical spacecraft launched a year apart to make first crossing of asteroid belt and first examination of Jupiter. Pioneer 10 launched on March 2, 1972 and reached Jupiter on December 3, 1973 to make 81 000-mile (130 330 km)-high flyby. Both will leave solar system, the first craft ever to do so. Spacecraft were also launched at highest speed ever achieved for a man-made object of 31 122 mph (50 075 km/h) by use of additional stage on launcher. Prime contractor: TRW
Launch vehicle: Atlas-Centaur plus TE-M-364-4 stage

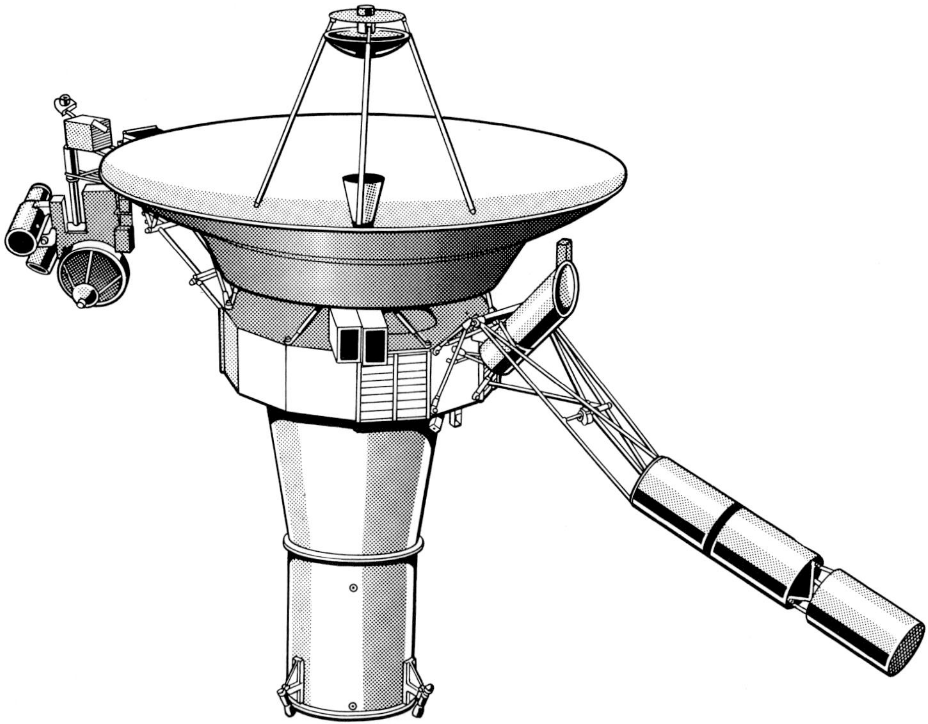

MARINER-JUPITER/SATURN

Role: Planetary fly-by
Data: Mariner-Jupiter/Saturn (early configuration)
Height: About 8 ft 4 in (2.5 m). With propulsion module 16 ft (4.9 m)
Diameter: Main body about 6 ft (1.8 m). Antenna 12 ft (3.6 m)
Launch weight: spacecraft 1 600 lb (726 kg). Propulsion module 2 600 lb (1 179 kg)
Spacecraft: Based on standard three-axis stabilised Mariner design but with 12-sided main body. Instrument bays fitted to sides surround single tank of hydrazine fuel for attitude control. Course-correction thrusters attached to aft rim of main body. High-gain dish antenna has low-gain antenna suspended at apex of feed reflector structure. Power is supplied by three radioisotope thermoelectric generators extended on single boom on opposite side of main body to scan platform. Propulsion module attached to base of main body has 16 000 lb (711.7 kN)-thrust solid fuel motor which is jettisoned after boosting spacecraft to 32 000 mph (51 500 km/h) trajectory. Three thrusters at tip of module provide stabilisation during burn
Payload: Total of 11 experiments (including radio signal investigations) weighing 190 lb (86 kg). Scan platform carries five experiments comprising high and medium resolution TV cameras, ultra-violet photometer, ultra-violet spectrometer, infra-red interferometer spectrometer and photopolarimeter. Platform mount structure carries plasma sensor and cosmic ray and low-energy charged-particle sensor. Twin whip-antennae for planetary radio astronomy and a boom carrying two magnetometers are deployed from base of RTG boom. New technique using high-frequency, high-power radio, allows instant transmission of TV pictures
Mission: Mariner 11 and Mariner 12 sister craft planned to reach Jupiter about a month apart in early 1979 and use gravitational swingby technique to reach Saturn in mid-1981. Mission includes mapping of at least ten of the 24 satellites of the two planets as well as planets themselves. Both spacecraft will leave the solar system
Programme: Two similar spacecraft in Mariner series are scheduled for launch in August and September 1977 in first US directly planned attempt at Saturn flyby. Mission replaces cancelled Grand Tour scheme and takes advantage of planetary line up to achieve short duration mission
Launch vehicle: Titan IIIE-Centaur-Burner II

(USA — Nasa) PIONEER VENUS ORBITER

Role: Planetary orbital survey
Data: Pioneer Venus 1978 orbiter (early configuration)
Height: Overall about 14 ft (4.3 m). Main body about 4 ft (1.2 m)
Diameter: about 9 ft (2.7 m)
Launch weight: 1 100 lb (499 kg)
Spacecraft: Drum-shaped main body is same as multiprobe main bus and is spin-stabilised with three-quarters of circumference covered with solar cells. Also carries liquid-fuel orbit insertion motor with nozzle protruding from base plus antennae assembly mounted on top. Design life is 225 days (one Venus year) after orbital entry
Payload: Selection of orbital survey instruments expected to include magnetometer, electron temperature probe, neutral mass and ultra-violet spectrometers, infra-red radiometer, radar altimeter and solar wind analyser plus X-band transmitter for occultation experiments
Mission: Spacecraft is planned to spend at least one Venus year operating in a highly eccentric orbit of 125 miles (200 km) to 37 300 miles (60 000 km) inclined at about 70 degrees. It is intended to provide radar mapping of surface as well as atmospheric and ionospheric studies
Programme: Pioneer Venus orbiter is single-spacecraft programme tentatively scheduled for launch in May 1978 to arrive in December. Prime contractor: Hughes
Launch vehicle: Atlas-Centaur

PIONEER VENUS PROBE

Role: Planetary atmospheric analysis
Data: Pioneer Venus 1978 multiprobe (early configuration)
Height: Overall about 8 ft (2.4 m). Main body about 4 ft (1.2 m)
Diameter: About 9 ft (2.7 m). Large probe about 6 ft (1.8 m), small probes about 3 ft (0.9 m) each across heatshield
Launch weight: Overall 1 800 lb (816 kg). Main body 600 lb (272 kg). Large probe 600 lb (272 kg), small probes 160 lb (73 kg) each
Spacecraft: Drum-shaped spacecraft is spin-stabilised and has body-mounted solar array covering three-quarters of circumference. Main body serves as carrier for one large and three small atmospheric entry probes attached on top. Probes are spherical pressure vessels enclosed in conical heatshields for aerodynamic breaking. Large probe also has parachute
Payload: Bus carries four experiments to measure upper atmosphere during entry. The large probe carries about ten experiments to measure atmospheric structure and composition and clouds plus complementary data for small probes which are designed to provide a three-dimensional picture of dynamics of lower atmosphere
Mission: Large probe separates 20 days and small probes 16 days before arrival. During 90-min descent, large probe jettisons heatshield at 42 miles (70 km) and parachute at 27 miles (44 km) giving freefall impact. Small probes make widely dispersed entries, free-falling to surface after heatshield jettison. Descent time is about 60 min. The probes are all designed to survive right down to surface but are not equipped to provide surface data. Entry of unprotected main bus at 75 miles (120 km) altitude is delayed until completion of probe missions
Programme: Pioneer multiprobe is single-spacecraft programme planned for launch in August 1978 to arrive in December. Mission is designed to complement that of Pioneer Venus orbiter scheduled to reach Venus six days earlier. Prime contractor: Hughes
Launch vehicle: Atlas-Centaur

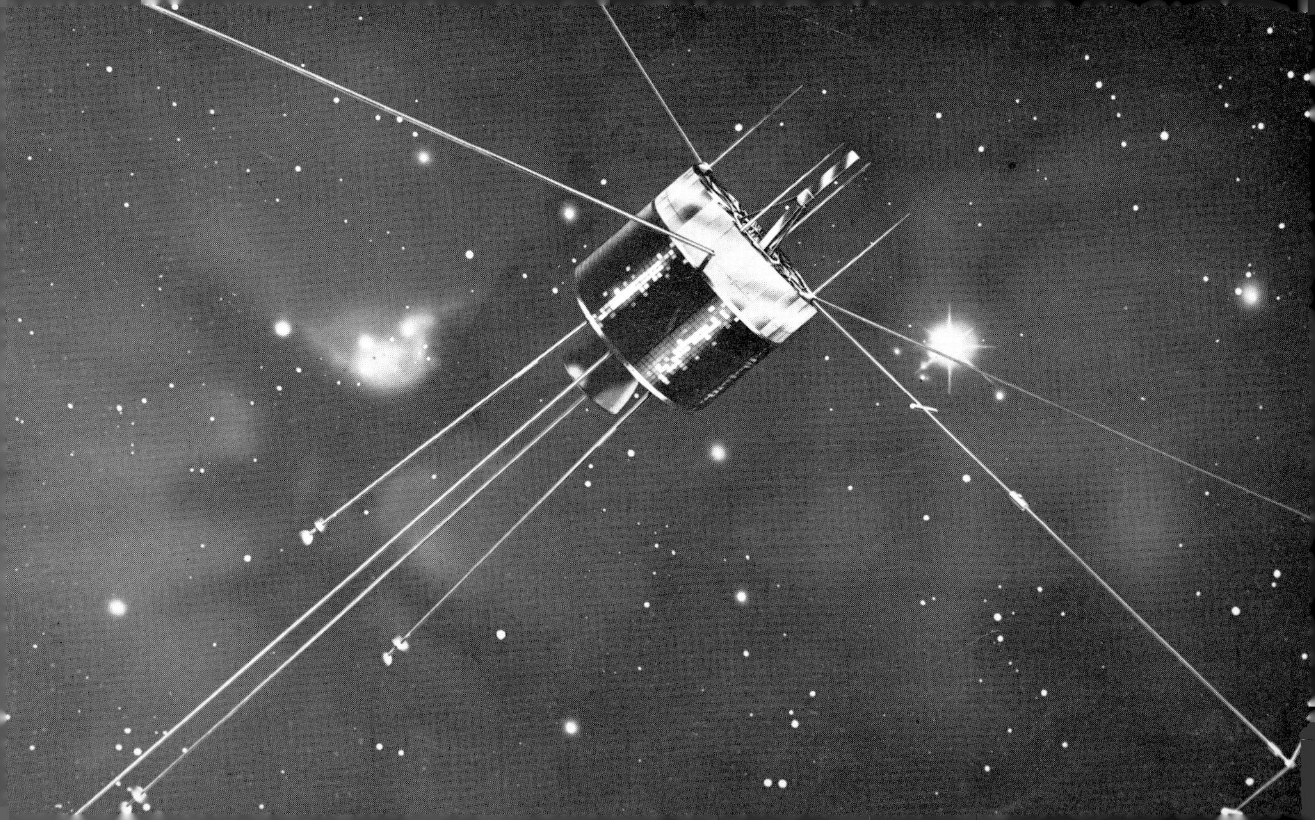

(Esro) GEOS

Role: Space science
Data: Geos
Height: 3 ft 9 in (1.1 m)
Diameter: 5 ft 4 in (1.62 m)
Orbital mass: 396 lb (180 kg). Launch weight: 1 195 lb (542 kg)
Spacecraft: Cylindrical spacecraft spin-stablised at 10 rpm in geostationary orbit. Body covered in solar cells generating 90 W. Projections include four magnetometer booms and various antennae, plus nozzle for apogee motor. Design life is two years
Payload: Nine experiments weighing only 70 lb (31 kg) to measure particle flux and electrical and magnetic fields at a constant distance from the Earth
Programme: Single spacecraft, scheduled for launch in mid-1976, is Esro's first geostationary scientific spacecraft. Prime contractor: BAC (for Star consortium)
Launch vehicle: Delta

(Esro)

HIGHLY ECCENTRIC ORBIT SATELLITE

Role: Space science
Data: Heos A2
Height: 2 ft 7 in (0.79 m). Overall: 7 ft 10 in (2.4 m)
Diameter: 4 ft 4 in (1.3 m)
Orbital mass: 258 lb (117 kg)
Spacecraft: 16-sided cylindrical spacecraft is spin-stablised at 10 rpm. Body-mounted solar array generates 55 W. Cold gas system provides reorientation of spin axis. Tripod projecting from top carries antennae and magnetometer boom. Design life is 18 months
Payload: Seven experiments to investigate the interaction of the solar particles and the Earth's magnetosphere at high latitudes
Programme: Heos A2 is second of two similar spacecraft launched in December 1968 and January 1972 into highly eccentric orbits taking them more than half way to the Moon's orbit. Heos 1 operated for five years and Heos A2 was still operating at end of 1974. Prime contractor: MBB
Launch vehicle: LTTA Delta

(Esro)

Role: Space science
Data: Esro 4
Height: 35 in (0.9 m)
Diameter: 30 in (0.76 m). With booms: 10 ft 9 in (3.5 m)
Orbital mass: 249 lb (113 kg)
Spacecraft: Cylindrical spacecraft is spin-stabilised at 65 rpm. Body-mounted solar array generates 60 W. Three experiment booms extend from base in orbit. Magnetic coil attitude control. Design life is 17 months
Payload: Five experiments to study the ionosphere, the near-Earth regions of the magnetosphere and auroral and solar particles, plus one engineering experiment (horizon sensor)
Programme: Esro 4 is individual spacecraft under Esro designation that has covered three previous scientific spacecraft. Esro 2 (Iris) was astronomical satellite launched in May 1968, Esro 1A (Aurorae) and Esro 1B (Boreas) were also ionospheric monitors launched in October 1968 and October 1969. Esro 4 launched in November 1972. Prime contractor: HSD
Launch vehicle: Scout

(Esro) TD-1A

Role: Stellar astronomy
Data: TD-1A
Height: 7 ft 1 in (2.16 m)
Width: 3 ft 3 in (1.0 m) square. Array span: 15 ft (4.55 m)
Orbital mass: 1 038 lb (472 kg)
Spacecraft: Rectangular body comprises a large lower section containing spacecraft systems and a small upper section containing experiments. Pair of solar arrays generates 330 W. Three-axis stabilisation provided by inertial wheels and argon gas jets (first European three-axis stabilised craft). Design life is six months
Payload: Seven experiments weighing 320 lb (145 kg) include broadband and selective ultra-violet telescopes, atomic particle counter, X-ray spectrometer and gamma radiometer, plus solar gamma radiometer and solar charged particle counter
Programme: Single spacecraft launched into 330-mile (531 km), near-polar orbit in March 1972. Second spacecraft cancelled. Sun-synchronous orbit maintained for six months and spacecraft was then spin-stabilised and shut down for four month eclipse period. Successfully reactivated for further six months. Prime contractor: Matra
Launch Vehicle: Thor Delta N (two-stage)

COS B

Role: Stellar astronomy
Data: Cos B
Height: 3 ft 11 in (1.2 m). With antennae: 5 ft 7 in (1.71 m)
Diameter: 4 ft 7 in (1.4 m)
Orbital mass: 616 lb (280 kg)
Spacecraft: Drum-shaped satellite spin-stabilised at 10 rpm. Body-mounted solar array generates 135 W. Cold-gas attitude control. Design life is two years

Payload: First single-experimental Esro payload is 265 lb (120 kg) gamma-ray spark chamber to extend TD-1A studies of stellar gamma radiation (chamber is four times the size of that in TD-1A). Also to complement Ariel 5 and ANS X-ray observations
Programme: Celestial Observation Satellite B is single spacecraft programme launched in August 1975 into highly eccentric polar orbit. Prime contractor: MBB (for Cesar consortium)
Launch vehicle: Delta

(ESA) EXOSAT

Role: Stellar astronomy
Data: Exosat
Orbital mass: 550 lb (250 kg)
Spacecraft: Drum-shaped spacecraft has single deployable solar array generating 100 W. Three-axis stabilisation supported by cold-gas jets, with hydrazine thrusters providing orbit control. Design life is two years
Payload: X-ray astronomy payload comprises two experiments for medium and low energy observations. Provides for location of X-ray sources by lunar occultation plus extended area study
Programme: Single spacecraft is scheduled for launch into highly eccentric orbit in 1979. Will extend studies of Ariel 5 and ANS.
Launch vehicle: Delta

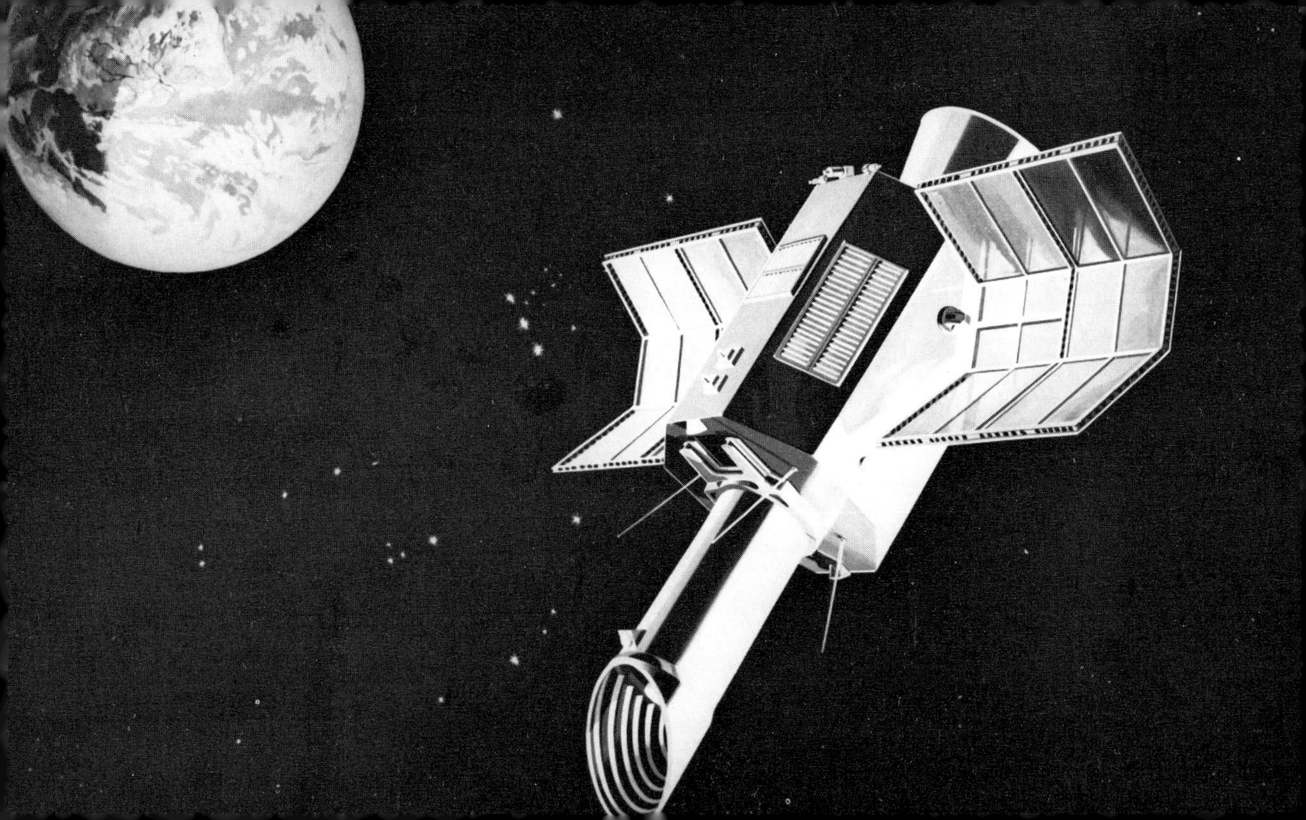

(Nasa/ESA/UK) INTERNATIONAL ULTRAVIOLET EXPLORER

Role: Stellar astronomy
Data: IUE
Orbital mass: 842 lb (382 kg). Launch weight 1 416 lb (642 kg)
Spacecraft: Octagonal cylinder has electronics and systems arranged in bays around central core containing telescope and apogee motor which project from opposite ends. Two deployable solar arrays generate 150 W. Spacecraft is gyrostabilised to 1 arc sec accuracy with cold gas thrusters providing station keeping and momentum dumping facility
Payload: Single experiment is 17.7 in (45 cm) Cassegrain telescope for ultra-violet spectroscopy
Programme: Only one spacecraft planned and this is intended to extend TD-1A studies and to bridge gap between Nasa's OAO and LST series. UK providing imaging tubes for spectrograph and ESA building solar arrays and European ground station in return for 30 per cent of satellite time plus full data exchange. IUE is due for launch in 1977. Prime contractor: Goddard Space Flight Centre
Launch vehicle: Delta

(ESA) INTERNATIONAL SUN-EARTH EXPLORER-B

Role: Space science
Data: ISEE-B
Height: 3 ft 3 in (1 m)
Diameter: 3 ft 6 in (1.1 m)
Orbital mass: 315 lb (143 kg)
Spacecraft: Cylindrical spacecraft has solar cells generating 80W covering body except for 8 in (22 cm) band at base. Deployable appendages include three 6 ft 6 in (2.0 m) long experiment booms and a pair of wire antennae measuring 98 ft (3.0 m) tip-to-tip. Cold-gas attitude control system also provides pulse firing at perigee to vary distance from ISEE-A. Spacecraft is spin-stabilised at 19.8 rpm. Design life is three years
Payload: Eight experiments weighing 28 lb (13 kg) comprise six particle investigations, an experiment to measure electric and magnetic waves and a magnetometer. Two experiments from Europe, remainder from USA
Programme: ISEE-B is "Daughter" spacecraft due to be launched with US-built ISEE-A "Mother" spacecraft (detailed separately) in October 1977. Programme is devoted to study of the Earth's magnetosphere in conjunction with Geos (ESA). Prime contractor: Dornier (for Star consortium)
Launch vehicle: Delta 2914

STARLETTE

(France — CNES)

Role: Geodesy
Data: Starlette
Diameter: 10.2 in (0.26 m)
Orbital mass: 110 lb (50 kg)
Spacecraft: Completely passive spacecraft is solid sphere with core of Uranium 235 to increase mass and surface enbedded with 60 laser relfectors
Programme: Single spacecraft launched in Feburary 1975 provides topographical data through laser telemetry and photography against stellar background over period of several years. Continues French geodetic studies initiated with three D-1 satellites in 1966-67 and Peole technology satellite (December 1970). Also served to qualify launch vehicle. Prime contractor: CNES
Launch vehicle: Diamant B-P4

D-5

Role: Satellite engineering
Data: D-5A (Pollux), D-5B (Castor)
Height: D-5A 1 ft 10 in (0.55 m). D-5B 2 ft 4 in (0.7 m)
Diameter: D-5A 2 ft (0.6 m). D-5B 2 ft 7 in (0.8 m)
Orbital mass: D-5A 79 lb (36 kg). D-5B 167 lb (76 kg)
Spacecraft: D-5A is polygon of 18 sides with solar cells covering alternate sides generating 5 W. Spin-stabilised at 10 rpm. Design life is seven months. D-5B is 26-sided polygon covered with solar cells generating 30 W Spacecraft is unstabilised. Design life is 5-6 years.
Payload: D.5A experimental hydrazine microthruster of 9 oz (2.5 N) thrust. D.5B testing micro accelerometer, measuring micrometeorites and assisting laser ranging experiments
Programme: Related pair of technology satellites were launched in May 1975 to replace indentical pair lost in launch failure in May 1973. Spacecraft are fourth and fifth French technology satellites following Sret 1 and 2 (both Soviet launched) and Peole. Prime contractor: Aerospatiale
Launch vehicle: Diamant B-P4

Role: Astronomy
Data: D-2B
Height: 2 ft 7 in (0.8 m)
Diameter: 2 ft 3 in (o.7 m). Array span: 8 ft 6 in (2.6 m)
Orbital mass: 242 lb (110 kg)
Spacecraft: Cylindrical main body has four solar panels hinged to midsection and generating 55 W. Spacecraft is spin-stabilised at 1 rpm and has cold-gas attitude control system. Upper section is experiment compartment with sensor apertures and is despun. Design life is eight months
Payload: Four experiments weighing 72 lb (33 kg) to study ultra-violet radiation (solar far-UV and stellar UV photometry, distribution and spectrophotometry)
Programme: D-2B, launched in September 1975 is third in D-2 series and seventh French scientific spacecraft since December 1965. Preceded by D-2A Tournesol launched in April 1971 and D-2A Polaire lost in launch failure in December 1971. Previous programmes are FR-1 (ionospheric) and D-1 (Diamant) series of geodatic satellites (D-1A, D-1C, D-1D). D-2B is last planned independent scientific satellite. Prime contracter: Matra
Launch vehicle: Diamant B-P4

HELIOS

(W. Germany)

Role: Solar astronomy
Data: Helios
Height: 7 ft (2.15 m). Overall 13 ft 10 in (4.21 m)
Diameter: 9 ft 1 in (2.7 m)
Orbital mass: 781 lb (355 kg)
Spacecraft: Spool-shaped spacecraft is spin-stabilised at 60 rpm. Strips of solar cells distributed around highly reflective body generate 240 W. Body design minimises surface exposure to solar heating. Design life is 18 months
Payload: Eleven experiments to study interplanetary magnetic effects cosmic dust and relectivity effects as well as measuring cosmic radiation, solar wind and solar X-rays. Three experiments provided by USA and remainder by Germany. Deployable appendages include 10 ft 6 in (3.2 m) magnetometer boom, 52 ft 6 in (16 m) antenna
Programme: Two spacecraft in German-US co-operative excercise will make closest ever approaches to the Sun, orbiting the Sun once every 192 days at a distance of about 0.3 AU (about 45 million km). First craft, Helios 1, launched in December 1974. Helios B due for launch in 1976. Prime contractor: MBB
Launch vehicle: Titan IIIE-Centaur

AEROS

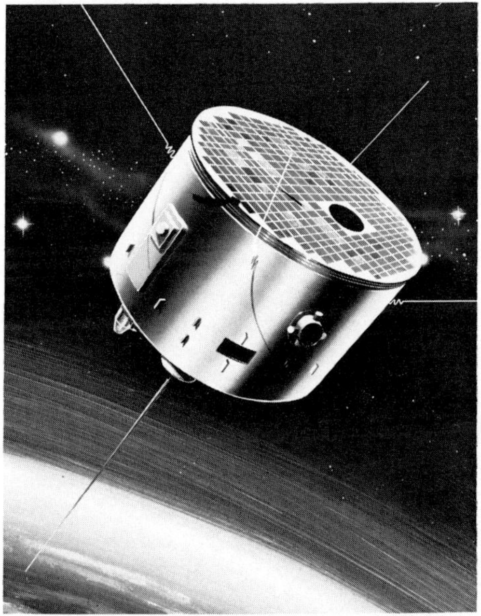

Role: Space science
Data: Aeros B
Height: 28 in (71 cm)
Diameter: 36 in (91 cm)
Orbital mass: 280 lb (127 kg)
Spacecraft: Main body is drum-shaped with slightly conical base from which extends a 71 in (1.8 m) impedance probe antenna. Top of drum covered in solar cells. Spacecraft is spin-stabilised with magnetic-coil attitude control and a pair of hydrazine-fuelled jets providing minor position adjustments. Design life is six months
Payload: Five experiments (one from USA) to study ionosphere and solar ultra-violet radiation effects
Programme: Initially Aeros single-spacecraft programme, launched in December 1972, extended to a second almost identical spacecraft Aeros B, launched in July 1974. Both placed in circular polar orbit. Prime contractor: Dornier
Launch vehicle: Scout

(India — Isro)

INDIAN SCIENTIFIC SATELLITE

Role: Space science
Data: ISS Aryabhata
Diameter: 4 ft 7 in (1.4 m)
Orbital mass: 727 lb (330 kg)
Spacecraft: 26-sided polygon with solar cells on all faces except top and bottom. Four small antennae project from middle of spin-stabilised spacecraft
Payload: Mixed payload of four experiments to study stellar X-ray emmissions, solar physics and ionospheric properties
Programme: First Indian national satellite launched by USSR in April 1975. All signals lost after five days and back-up now planned for launch by USSR in 1977. ISS originally sized smaller for Indian launcher but enlarged and brought forward after Soviet launch offer. Indian national SLV-3 launcher due to enter service in 1978-80 with launch of second Indian satellite weighing 88 lb (40 kg). To be followed by 1983-84 launching of 1 750 lb (800 kg) Indian National Satellite (Insat) into geostationary transfer orbit by more advanced SLV-4. Prime contractor: Isro
Launch vehicle: B-1

SAN MARCO

Role: Space science and engineering
Data: San Marco 4
Diameter: 2 ft 6 in. (0.76 m)
Orbital mass: 363 lb (165 kg)
Spacecraft: Spherical structure has four 19 in (48 cm) antennae on top and double band of solar panels around equator. Spacecraft is spin-stabilised at 6 rpm
Payload: Two US-supplied experiments to study constitution of upper atmosphere and Italian experiment measuring density by atmospheric braking
Programme: Latest in series of four US-Italian co-operative spacecraft launched between December 1964 and February 1974 for aeronomy and (in first two) ionospheric research and to train Italian launch crews. First launch from Wallops Island and remainder from San Marco launch platform. Prime contractor: CRA (Rome University Aerospace Research Centre)
Launch vehicle: Scout

(Japan — Nasda)

IONOSPHERE SOUNDING SATELLITE

Role: Communications technology
Data: ISS
Height: 32 in (0.82 m)
Diameter: 37 in (0.94 m)
Orbital mass: 298 lb (135 kg)
Spacecraft: Spin-stabilised spacecraft is cylinder covered with solar cells on curved surface
Payload: Radio transmitters for experiments on interaction of radio signals with inonosphere to map worldwide distribution of critical frequencies
Programme: ISS is only spacecraft within specific programme but is also the second of three development payloads for the N launcher. ISS scheduled for launch in early 1976 into 620 mile (1 000 km) orbit similar to that of ETS (Engineering Test Satellite)-1 launched in September 1975. ETS-2 scheduled for launch into geostationary orbit in 1977 (see ECS). Prime contractor: Mitsubishi
Launch vehicle: N

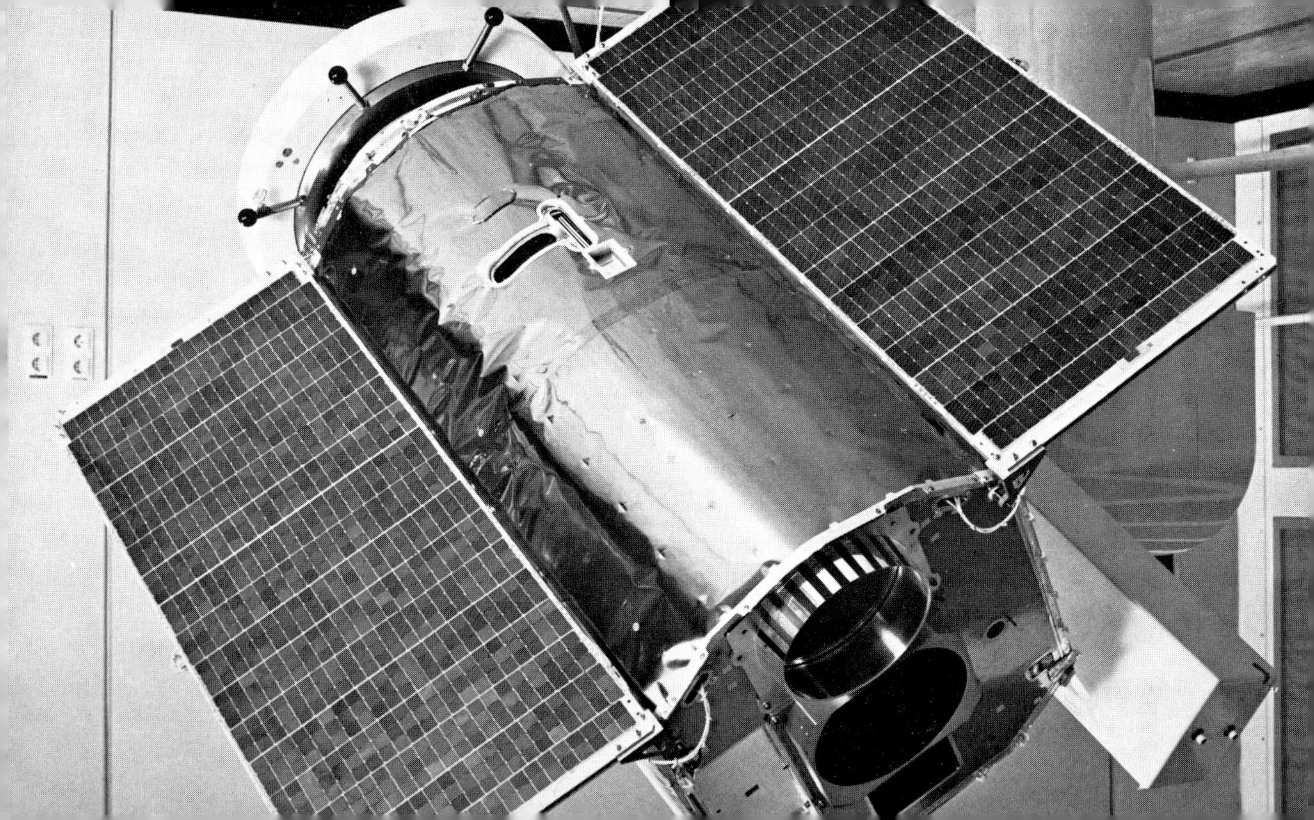

(Netherlands)

ASTRONOMICAL NETHERLANDS SATELLITE

Role: Stellar astronomy
Data: ANS
Height: 48 in (1.23 m)
Depth: 29 in (0.73 m)
Width: 24 in (0.61 m). Array span: 57 in (1.44 m)
Orbital mass: 271 lb (123 kg)
Spacecraft: Three-axis stabilised satellite with pair of deployable solar arrays. Control provided by magnetic coil and momentum wheels. 12 h data storage capability. Design life is six months
Payload: Three experiments; hard X-ray counter (US-supplied), soft X-ray counter and Cassegrain telescope for ultra-violet radiation from young, hot stars
Programme: Single spacecraft launched in August 1974 into sun-synchronous polar orbit. Launcher fault prevented attainment of planned circular orbit and spacecraft operated satisfactorily from a high eccentric orbit. Second ANS now planned. Prime contractor: Phillips/Fokker VFW consortium
Launch vehicle: Scout

INTASAT

(Spain)

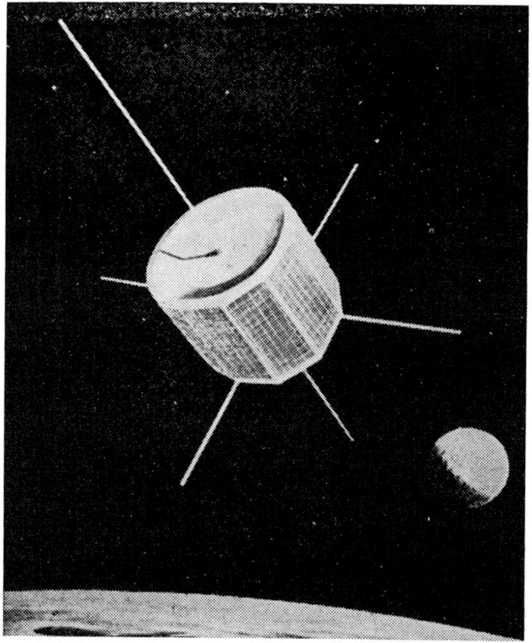

Role: Space science
Data: Intasat
Height: 17.75 in (45 cm)
Diameter: 17.5 in (44.5 cm)
Orbital mass: 54 lb (24.5 kg)
Spacecraft: Spin-stabilised spacecraft is 12-sided cylinder with two aerials extending along spin axis and four projecting laterally from top. Body-mounted solar array generates 2.77 W. Design life is two years
Payload: Radio equipment to relay to Earth signals from high altitude balloons. Balloon transmits polarised signals linearly on 40 MHz and 41 MHz to measure density of ionosphere
Programme: Single spacecraft, launched into Sun-synchronous polar orbit in November 1974 in triple launch with NOAA 4 and Oscar 7, is Spain's first satellite. Intended to develop space engineering capability as much as serve scientific purpose. Also designed as standard platform for experiments up to 6½ lb (3 kg) in weight. Prime contractor: INTA
Launch vehicle: Delta

UK-6

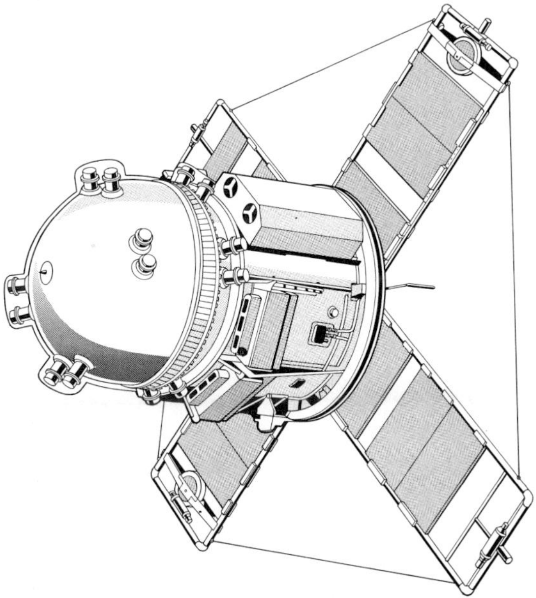

Role: Stellar astronomy
Data: UK 6 (Ariel 6)
Height: 4 ft 3 in (1.3 m)
Diameter: 3 ft 2 in (0.96 m)
Orbital mass: 331 lb (150 kg)
Spacecraft: Cylindrical main body has four solar arrays deployed from top generating at least 72 W. Spacecraft is spin-stabilised at 65 rpm. Reaction control provided wholly by magnetorquer. Design life is two years
Payload: Three experiments weighing 340 lb (154 kg) include two X-ray experiments (one built in collaboration with USA) attached to sides plus spherical cosmic ray experiment fitted to base of main body
Programme: Latest spacecraft in UK/Ariel series is scheduled for launch in late 1977. Prime contractor: Marconi
Launch vehicle: Scout

X SERIES

Role: Space engineering
Data: X-4 Miranda
Height: 2 ft 9 in (0.83 m)
Width: 2 ft 2 in (0.66 m). Array span: 8 ft 2 in (2.5 m)
Orbital mass: 206 lb (93 kg)
Spacecraft: Rectangular spacecraft has two deployable solar arrays at sides and four small antennae protruding from base. Design life is six months
Payload: Five technology experiments include three-axis attitude control, new solar cells, a star sensor and infra-red and albedo horizon sensors
Progamme: X series of technology satellites originally designed for launch by Black Arrow. X-3 Prospero, launched in October 1971, was first and only Black Arrow launch. X-4 launched by Nasa in March 1974. Original X-5 proposal has evolved into OTS. Prime contractor: (X-4) HSD
Launch vehicle: Scout

ARIEL

(UK)

Role: Space sciences
Data: Ariel 5
Height: 34 in (0.86 m). With antennae: 55 in (1.41 m)
Diameter: 38 in (0.96 m)
Orbital mass: 298 lb (135 kg)
Spacecraft: Cylindrical spacecraft (with 16 facets) is spin-stabilised at 10 rpm with solar cells covering most facets of body. Propane gas jets provide pointing control while drift is inhibited by magnetic torque. Design life is one year
Payload: Six experiments designed to locate and measure X-ray sources within and beyond our galaxy. Two instruments (one provided by the USA) look through side of spinning craft to monitor known X-ray sources and detect new ones. The other four instruments, directed through the top of the craft, can be pointed quickly at sources found by the side scanners to provide detailed investigation
Programme: Non-specific UK-USA co-operative programme of space sciences has involved five assorted spacecraft launched since April 1962. UK-6 under development for launch in 1977. First two were US-built and carried some UK experiments. Missions: Ariel 1 — ionosphere; Ariel 2 — ozone, micrometeoroids and galactic radio noise; Ariel 3 — upper atmosphere and radio noise data; Ariel 4 — ionosphere study. Ariel 5 launched in October 1974 from San Marco into near equatorial orbit to complement US missions SAS-A and OAO-3. Prime contractor: Marconi
Launch vehicle: Scout

(USA — Nasa)

GEOS

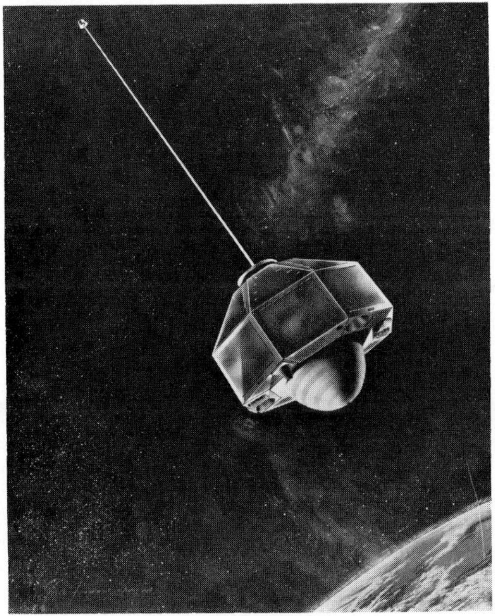

Role: Geodesy/oceanography
Data: Geos 3
Height: 2 ft 8 in (0.81 m)
Diameter: 4 ft 5 in (1.32 m)
Orbital mass: 750 lb (340 kg)
Spacecraft: Eight-sided structure covered with solar cells is surmounted by truncate cone from which extends 22 ft (6.0 m)-long scissor-type boom carrying 100 lb (45 kg) end mass for gravity gradient stabilisation. Momentum wheel extends this to three-axis stabilisation. Experiment platform on base carries antennae and laser reflectors. Design life is one year
Payload: Five instruments provide readings for 13 different measurements studies of solid Earth physics and oceanography plus unprecedented high precision satellite tracking by multiple techniques including laser and via ATS-6. Instruments include radar altimeter (first ever carried on an unmanned spacecraft), two C-band transponders, one S-band transponder (for ATS-6 tracking), laser reflectors and a radio dopplar system
Programme: Geodynamics Experimental Ocean Satellite (Geos)-3 is latest in series of six different satellites in US National Geodetic Satellite Programme launched between October 1964 and March 1975. Initiated by Beacon Explorer A and B (Explorers 22 and 27) followed by Geos 1 and 2 (Explorers 29 and 36) plus Pageos 1 (Passive Geos) balloon reflector. Programme also includes use of radar altimeter aboard Skylab and Lageos reflecting satellite (detailed separately Geos-3 objective is precise measurement of ocean surface topography, and sea state. Prime contractor: Applied Physics Laboratories
Launch vehicle: Two-stage Delta

LAGEOS

(USA — Nasa)

Role: Geodesy
Data: Lageos
Diameter: 2 ft (0.6 m)
Orbital mass: 906 lb (411 kg)
Spacecraft: Completely passive spacecraft is solid sphere covered with 426 slightly recessed relecters each 1.5 in (3.8 cm) in diameter. Sphere consists of cylindrical brass core enclosed by two alluminium alloy fitted hemispheres and with a single central bolt securing the assembly. Extra stage (TE-M-604) attached to spacecraft and jettisoned after burn out
Programme: Laser Geodynamic Satellite (Lageos) is single spacecraft scheduled for launch in March 1976 into high permanent orbit (carries message for any future interceptor). Mobile laser ranging stations will move through 14 countries throughout planned eight-year programme taking measurements for study of continental drift, tides, fault motions. Planned completion by 1978 of full network of seven stations will provide measurement accuracy of 2 cm. Prime contractor: Marshall Space Flight Centre
Launch vehcle: Delta

(USA — Nasa)

ORBITING GEOPHYSICAL OBSERVATORY

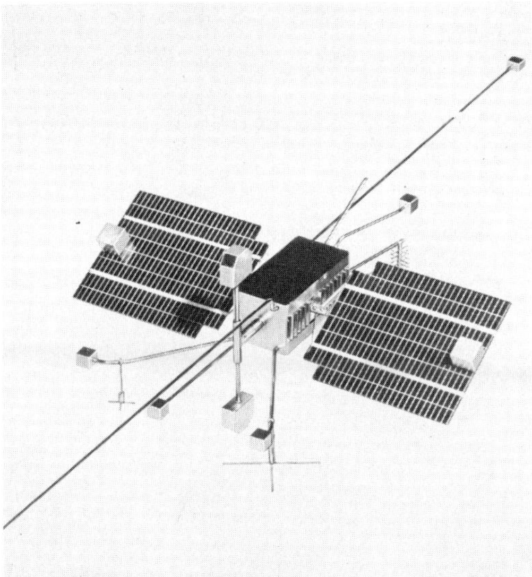

Role: Geophysical investigation
Data: OGO-6
Length: 6 ft (1.8 m). With booms: 49 ft (14.9 m)
Width: 3 ft (0.9 m) square. Array span: 20 ft 6 in (6.2 m)
Orbital mass: 1 393 lb (632 kg)
Spacecraft: Rectangular main body carries two deployable 22 ft (6.7 m) long and four 4 ft (1.25 m)-long experiment booms and two 6 ft x 8.5 ft (1.8 m x 2.6 m) solar panels generating 560 W. Three-axis stabilisation provided by gas-jet and flywheel system
Payload: 371 lb (168 kg) multipurpose payload comprises 25 experiments, seven atmospheric and ionospheric, three for solar radiation, ten for airglow and auroral studies and five for magnetic and electrical field studies. Two experiments failed to function
Programme: Six spacecraft carried total of 130 experiments into a variety of Earth orbits — including eccentric orbit (EGO) and polar orbit (POGO) — between September 1964 and June 1969. Highly successful programme claimed to have provided more data than all other scientific satellites combined, including a number of first discoveries. Prime contractor: TRW
Launch vehicle: Thorad-Agena D (POGOs — Atlas-Agena)

ATMOSPHERE EXPLORER

(USA — Nasa)

Role: Aeronomy
Data: AE-C (Explorer 51)
Height: 3 ft 9 in (1.15 m)
Diameter: 4 ft 5 in (1.35 m)
Orbital mass: 1 450 lb (680 kg)
Spacecraft: Cylindrical spacecraft with top and sides covered with solar cells generating 100 W. Optional three-axis or spin-stabilisation variable from 0.5 — 10 rpm using momentum wheel, magnetic bias coil and nutation dampers. 370 lb (168 kg) hydrazine supply provides 200 miles (320 km) orbit adjustment capability for maintaining orbit at minimal perigee. 21 miles (32 km) of this is required during the one-year design life period
Payload: 14 experiments to obtain data on the composition of the upper atmosphere above 72 miles (115 km) altitude and its interactions with incoming radiation
Programme: Launched in December 1973, AE-C is first of three second-generation Atmosphere Explorers. Predecessors were Explorers 17 and 32 launched in April 1963 and May 1966. Two more (D and E) were due for launch in September and December 1975. Prime contractor: RCA
Launch vehicle: Two-stage Delta

(USA/Europe — Nasa/ESA) **INTERNATIONAL SUN-EARTH EXPLORER PROGRAMME**

Role: Space science
Data: ISEE-A
Height: 5 ft 3 in (1.6 m). Overall 13 ft 2 in (3.9 m)
Diameter: 5 ft 8 in (1.7 m)
Orbital mass: 749 lb (340 kg)
Spacecraft: Cylindrical body with 16 sides is divided into three horizontal bands with outer two covered by solar cells generating 150 W. Freon cold gas attitude-control system has thruster at ends of opposing 43 in (1.3 m) long booms. Two 12 ft (3.6 m) long experiment booms and two pairs of experiment antennae measuring 262 ft (80 m) and 702 ft (214 m) tip-to-tip, all projections deploying from centre band of body. Core tube is truncated cone providing 18 in (0.46 m) diameter ISEE-B mounting at top and 37 in (0.97 m) diameter launcher adapter at base. Also contains antenna mast deployed after A/B separation. Spacecraft is spin-stabilised at 19.7 rpm. Design life is three years
Payload: Package of 14 experiments weighing 78 lb (35 kg) has same experiments as ISEE-B plus reinforced wave measurement, two more particle detectors and two to measure isotopic composition of cosmic rays and solar particles
Programme: ISEE programme is Nasa/ESA co-operative study of Earth's magnetosphere using three co-ordinated spacecraft plus data from Geos. (Esro). Double launch into eccentric Earth orbit of ISEE-A (Mother) and -B (Daughter) is scheduled for October 1977. ISEE-C (Heliocentric) to follow in July 1978. ISEE-A and -C are similar, built by Nasa and based on IMP design (enlarged). ISEE-C will weigh 948 lb (430 kg) and have hydrazine ACS and 13 experiments. ISEE-B (detailed separately) is being provided by ESA. Prime contractor: (ISEE-A) Goddard Space Flight Centre
Launch vehicle: Delta 2914

APOLLO SUB-SATELLITE

(USA — Nasa)

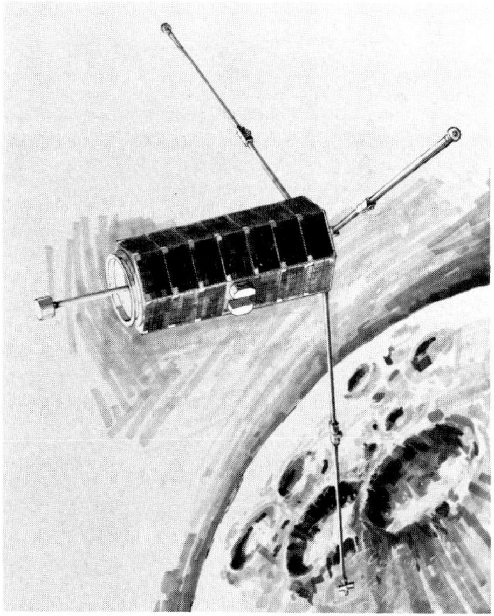

Role: Lunar orbital survey
Data: Particles and Fields satellite (Apollo 16)
Height: 31 in (79 cm)
Diameter: 14 in (35 cm)
Orbital mass: 80 lb (36 kg)
Spacecraft: Cylinder with six sides is spin-stabilised at 12 rpm. Body-mounted solar array generates 25 W. Design life is 1 year
Payload: S-band transponder of 1 W power used to discern effect of gravitational anomalies on spacecraft for mapping purposes. Also equipped to analyse particle content of near-lunar space and interaction of magnetic fields of Moon and Earth
Programme: Apollo 16 P&F satellite injected into lunar orbit in April 1972 from Apollo CSM SIM bay just before departure. Early departure necessitated wrong orbit, reducing satellites orbital life to one month. Preceded by identical Apollo 15 P&F satellite left in August 1971
Prime contractor: TRW
Launch vehicle: Saturn 5/Apollo CSM

(USA — Nasa)

INTERPLANETARY MONITORING PLATFORM

Role: Space sciences
Data: IMP-J (Explorer 50)
Height: 5 ft 2 in (1.57 m)
Diameter: 4 ft 5 in (1.35 m)
Orbital mass: 818 lb (371 kg)
Spacecraft: Sixteen-sided, drum-shaped satellite is spin-stabilised at 23 rpm and uses cold-gas attitude control. Body surface divided horizontally into four tiers, three covered with solar arrays generating 110 W and the fourth carrying experiment booms and sensors. Two 10 ft (3.0 m) magnetometer booms and two 4 ft (1.3 m) attitude-control system booms deploy in orbit. Spacecraft divided into lower compartment containing solid-fuel kick motor and upper compartment containing experiments and control systems. Design life is one year
Payload: Twelve experiments designed to measure magnetic and electric fields, energetic particles and plasmas. Also, two technology experiments (new solar cells and new data multiplex unit)
Programme: IMP J is last in series of ten craft launched between November 1963 and October 1973 to cover 11-year cycle of solar activity. Initial programme involved series of seven smaller craft (Explorers 18, 21, 28, 33, 34, 35 and 41). Two were fitted with solid-fuel braking motors to place them in lunar orbit (Anchored IMP-AIMP). One succeeded (Explorer 35) Remainder placed in different highly eccentric Earth orbits, as were second series of three (Explorers 43, 47 and 50). Prime contractor: (IMP-J) Goddard Space Flight Research Centre
Launch vehicle: Various Delta models

INJUN

(USA — DOD/Nasa)

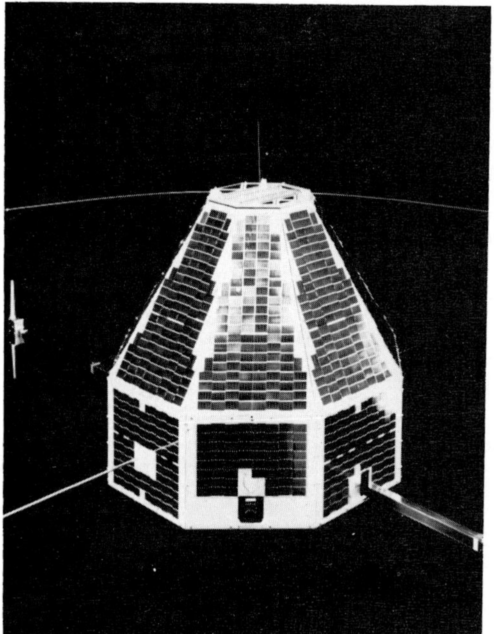

Role: Space science
Data: Hawkeye 1 (Explorer 52)
Height: 2 ft 6 in (0.75 m)
Diameter: 2 ft 6 in (0.75 m)
Orbital mass: 59 lb (26.6 kg)
Spacecraft: Eight-sided drum surmounted by truncated cone is covered on all sides and base by solar cells and spin-stabilised at 5-10 rpm. Design life is one year
Payload: Three experiments to investigate interaction of solar wind with Earth's magnetic field are a flux-gate magnetometer on a 5 ft (1.6 m) boom, a low frequency (VLF-ELF) electric field receiver using antennae spanning 150 ft (46 m) and another 5 ft boom plus an electron and proton analyser
Programme: Hawkeye 1 is first in new series under Injun programme of Explorers which now totals six varied spacecraft launched between June 1961 and June 1974 to study solar particles. Programme initiated by US Navy. Injun Explorers 25, 40 and 52 operated by Iowa University. Prime contractor: Iowa University
Launch vehicle: Scout E (five-stage)

(USA — Nasa/DOD)

SOLAR RADIATION SATELLITE

Role: Solar astronomy
Data: Solrad C (Explorer 44)
Height: 23 in (0.58 m)
Diameter: 30 in (0.76 m)
Orbital mass: 260 lb (118 kg)
Spacecraft: 12-sided cylinder with four small solar panels projecting from centre section. Spin-stabilised at 60 rpm. Design life is three years
Payload: 14 experiments to monitor continuously solar X-ray and ultra-violet emissions. Can also be redirected to record stellar emissions
Programme: Solrad C is third in a series of small, assorted satellites initiated for the International Quiet Sun Years of 1964/1965 by Solrad A (Explorer 30), launched in November 1965. Solrad B (Explorer 37) launched in March 1968. Co-operative programme with US Navy. Series also designated Solrad 8, 9 and 10 to link with preceeding military solar observatories in the OV-1 series. Prime contractor: US Navy Research Laboratory
Launch vehicle: Scout

RADIO ASTRONOMY EXPLORER

(USA — Nasa)

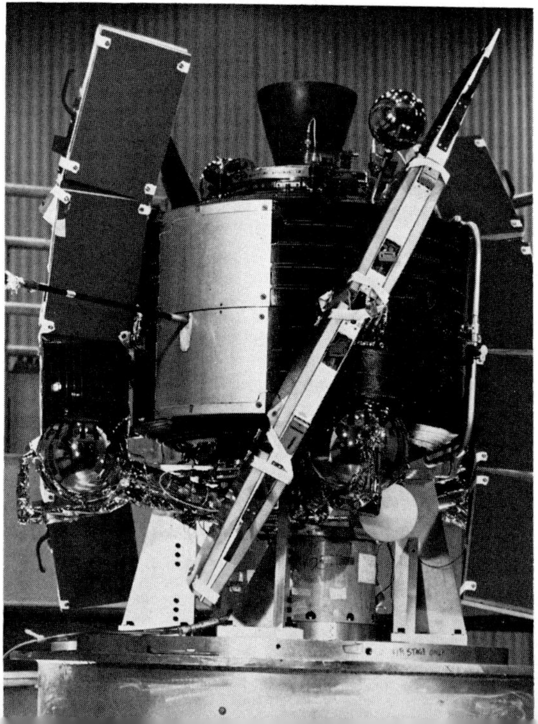

Role: Stellar and planetary astronomy
Data: RAE-B (Explorer 49)
Height: Main body 2 ft 7 in (0.79 m)
Diameter: 3 ft (0.92 m). Antennae span 1 300 ft (396 m)
Orbital mass: 442 lb (200 kg). Launch weight 730 lb (328 kg)
Spacecraft: Cylindrical main body has four 5 ft (1.5 m)-long canted solar panels on sides generating 38.5 W. Freon gas attitude-control thrusters on short booms control spin-stabilisation during translunar flight and three-axis stabilisation in orbit (RAE-A was gravity-gradient stabilised). Solid-fuel orbit insertion motor on top of main body and hydrazine-fuelled Velocity Control Propulsion System on base are jettisoned separately after lunar orbit is established. Design life is one year
Payload: Radio equipment and antennae to map planetary and stellar low frequency radio emissions. Main antennae comprise four 750ft (228 m)-long bimetallic strips which curl into half-inch tubes during slow deployment, forming "X" shape. Other appendages are a 120ft (37.0 m)-span dipole antenna and a 675 ft (200 m)-span libration damper boom
Programme: Two spacecraft programme initiated in July 1968 with launch of RAE-A (Explorer 38) into Earth orbit where it operated for four years. RAE-B placed in lunar orbit in June 1973 to use Moon as screen against terrestrial radio noise. Spacecraft was also last planned US space mission to the Moon. Prime contractor: Goddard Space Flight Centre
Launch vehicle: Delta 2913

(USA — Nasa)

SMALL ASTRONOMY SATELLITE

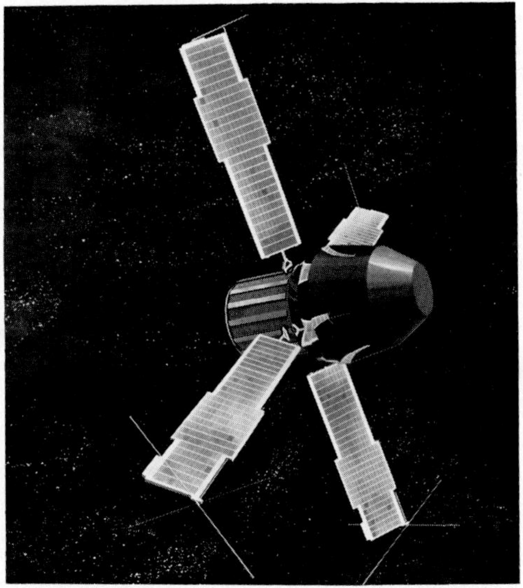

Role: Stellar astronomy
Data: SAS-B (Explorer 48)
Height: 4 ft 3 in (1.29 m)
Diameter: 1 ft 10 in (0.55 m). Array span: 13 ft (4.0 m)
Orbital mass: 410 lb (186 kg)
Spacecraft: Spin-stabilised satellite comprises cylindrical control section surmounted by bulbous experiment compartment and carrying four deployable solar arrays. Design life is 1 year
Payload: Single experiment is highly sensitive gamma-ray telescope to detect and pinpoint new radiation sources
Programme: SAS-B is second of three planned small astronomy satellites and was launced into equatorial orbit in November 1972. Power-supply fault caused shutdown after six months operation. SAS-A (Explorer 42), named Uhuru after launch in December 1970, carried X-ray spectrometer to make first detailed study of stellar X-ray sources, detecting 160 new sources. SAS-C (Explorer 52) launched in May 1975 with four-experiment X-ray payload. Prime contractor: Goddard Space Flight Centre
Launch vehicle: Scout

ORBITING SOLAR OBSERVATORY

Role: Solar astronomy
Data: OSO-7
Height: Wheel 9 in (22 cm). Sail 23 in (0.6 m). Overall 3 ft 1 in (0.92 m)
Diameter: 3 ft 8 in (1.13 m). Overall 7 ft 8 in (2.31 m)
Orbital mass: 1 400 lb (635 kg)
Spacecraft: Similar in appearance to previous OSOs but more than twice as heavy because of increased payload and support systems. Also, spin-stabilisation of 30 rpm is controlled by balance weights instead of momentum arms. Spacecraft comprises nine-sided base section (or "wheel") containing control systems and an upper despun "sail" carrying 38 W solar array. Sun-orientation is maintained by nitrogen-gas jets augmented by magnetic torquing coil for fuel conservation. Design life is six months

Payload: Six experiments primarily to study solar corona. Sail carries two experiments X-ray and extreme ultra-violet spectroheliographs and white light and extreme u-v coronagraphs which can be pointed at a number of large or small discretely variable points. Four experiments in the spinning base investigate high and low energy cosmic X-rays, hard solar X-rays and solar gamma rays
Programme: Nasa's first standard observatory spacecraft. Seven craft of variable payloads (plus one lost in launch failure) placed in 350-mile (563 km) near-circular orbit between March 1962 and September 1971. Objective was to cover peak of 11-year cycle of solar activity. OSO-5 (January 1969) still functioning. OSO-7 in eccentric orbit because of launcher separation fault. Prime contractor: Ball Brothers
Launch vehicle: LTTA Delta (double launch with small TETR satellite)

ADVANCED OSO

(USA — Nasa)

Role: Solar astronomy
Data: OSO-8
Height: Wheel 2 ft 4 in (0.72 m). Sail 7 ft 8 in (2.35 m)
Diameter: Wheel 5 ft (1.5 m). Sail 6 ft 10 in (2.1 m)
Orbital Mass: 2 346 lb (1 064 kg)
Spacecraft: Main body is drum-shaped "wheel" with three balance weight booms projecting from sides and spin-stabilised at 6 rpm. Attitude controlled by magnetic torque coils plus nitrogen-gas jets for spin-axis yaw and pneumatic system for spin and pitch. "Sail" is despun solar array generating 110 W and with Pointed Instrument Assembly inserted in base. Design life is one year
Payload: Total of eight experiments comprises two pointed instruments in sail — high resolution ultra-violet spectrometer and French-supplied chromosphere fine structure analyser — and remainder in wheel section. These include crystal spectrometer and polarimeter, mapping X-ray heliometer, soft X-ray background collimater, cosmic X-ray spectroscope, high energy celestial X-ray experiment and extreme u-v radiation experiment
Programme: Single spacecraft, launched in June 1975, is eighth in OSO series (see OSO) and twice as large as predecessor. Carries most elaborate selection of X-ray telescopes yet launched plus two principal instruments for u-v astronomy. Prime contractor: Hughes
Launch vehicle: Delta "Straight Eight" (two-stage)

(USA — Nasa)

ORBITING ASTRONOMICAL OBSERVATORY

Role: Stellar astronomy
Data: Copernicus (OAO-3)
Height: 9 ft 10 in (3.0 m). Overall: 16 ft 1 in (5.92 m)
Diameter: 6 ft 8 in (2.03 m). Array span: 21 ft 2 in (6.45 m)
Orbital mass: 4 900 lb (2 223 kg)
Spacecraft: Octagonal cylinder has equipment bays fitted to inside of walls enclosing a 4 ft (1.25 m) diameter core for experiments, plus 6 ft-long telescope sunshield on top. Two deployable solar arrays generate 1.6 KW. Three-axis stablisation uses four gimballed star trackers, and inertial reference unit and stabilising gyros to provide pointing accuracy of less than 1 second of arc — highest accuracy of any Nasa spacecraft. Design life is one year
Payload: Two experiments weighing 1 000 lb (454 kg). Main experiment is Princeton Experiment Package, a 10 ft (3 m) long ultra-violet telescope with a 32 in (81 cm) mirror to study young hot stars and interestellar gas and is fitted into central core. The University College of London Experiment Package comprises three small telescopes and an X-ray collimator to pinpoint X-ray sources and is carried in an upper bay of the main body
Programme: Standard platform for a variety of telescopes used in four missions between April 1966 and August 1972 to study interstellar absorption of common elements in interstellar gas and investigate ultra-violet radiation from young, hot stars. OAO-1 lost from power supply failure on second day and third craft (OAO-B) lost in launch failure. Performance and data from OAO-2 and -3 exceeded all expectations. Prime contractor: Grumman
Launch vehicle: Atlas-Centaur

NASA

LARGE SPACE TELESCOPE

Role: Stellar astronomy
Height: From 40 ft to 48 ft (12 m to 14.6 m)
Diameter: From 13 ft to 15 ft (4 m to 4.6 m)
Orbital mass: From 15 000 lb to 20 000 lb (6 800 kg to 9 070 kg)
Spacecraft: Cylindrical spacecraft comprises shielded telescope with light baffle projecting from viewing end and support systems module attached to base from which project two solar arrays. Three-axis stabilisation provides pointing accuracy of 0.01 arc sec
Payload: Telescope with 95 in (2.4 m) aperture is supported by selection of instruments for variety of astronomical studies. Instruments can be changed periodically. Images are televised for transmission.
Programme: Single multi-purpose optical telescope is scheduled for launch in late 1982 into 300 mile (500 km) orbit. LST will be serviced by Space Shuttle to provide indefinite lifetime and will be returned to Earth for periodic changes of instruments. Development planned to begin in 1977.

EXPERIMENTS
TELEMETRY
GAS MANAGEMENT
LIFE SUPPORT
N_2 GAS STORAGE
BIOSATELLITE

BIOSATELLITE

Role: Space science and engineering
Data: Biosatellite 3
Length: 6 ft 9 in (2.05 m)
Diameter: 4 ft 9 in (1.44 m)
Orbital mass: 1 536 lb (697 kg)
Spacecraft: Three segment spacecraft comprises 57 in (1.44 m)-diameter cylindrical adapter module. Forward end tapers to 40 in (1.01 m) interface with re-entry vehicle which encloses recovery capsule. Adapter contains spacecraft control subsystems including three stabilisation gyros and nitrogen-gas tanks for attitude control jets. Also contains life support system providing Earth-type atmosphere of 80 per cent nitrogen and 20 per cent oxygen at 14.7 lb/sq in (703 gm/cm²). Power supplied by fuel cell which also provides drinking water. Re-entry vehicle is 40 in (1.01 m)-diameter bowl-shaped glass fibre heat shield with phenolic nylon ablative covering and jettisonable solid-fuel retro-rocket of 1 000 lb (4.45 kN) thrust fitted to flat aft end. Recovery capsule is blunt aluminium cone with 9 cu ft (0.25 m³) internal volume
Payload: Single pigtail monkey fitted with sensors to provide 24 physical measurements constantly throughout 30-day flight
Programme: Biosatellite 3 is third in series of four planned space biology missions (reduced from six originally planned) launched between December 1966 and June 1969 and first to carry advanced organism. Also first US spacecraft ever to use Earth-type atmosphere. Mission terminated early after nine days when monkey became dehydrated and died 12 h after ocean recovery. First two flights carried plants and small animals. Fourth planned to repeat third. Prime contractor: General Electric
Launch vehicle: LTTA Delta N (two-stage)

(USA — Nasa)

SPACE ELECTRIC ROCKET TEST

Role: Spacecraft engineering
Data: Sert 2
Length: 20 ft 8 in (6.3 m)
Diameter: 5 ft (1.52 m)
Orbital mass: 3 100 lb (1 406 kg)
Spacecraft: Cylindrical spacecraft is Agena satellite equipped with pair of deployable solar arrays generating 1,471 W.
Payload: Spacecraft ring 40 in (1.0 m) long and 59 in (1.24 m) in diameter mounted on forward end of Agena contains experimental ion engines. Two hollow cathode ion bombardment engines of 0.006 lb (0.027 N) thrust and 29 lb (13 kg) of mercury propellent for each engine plus control systems. Engine exhaust velocity was 50,000 mph (80,000 km/h). Also had related engineering experiments
Programme: Sert 2, launched in February 1970, was first US orbital test of ion engines. Engines were intended to operate consecutively for six months each. First failed after five months and second after three months. Both failures due to electrical short in high voltage supply but fault subsequently corrected by manoeuvring spacecraft. Engines have been reactivated for numerous short runs since May 1974. Mission preceded by Sert 1 suborbital flight in July 1964 in which one engine failed but second provided more than 30 min operation. Prime contractor: Lockheed
Launch vehicle: Thorad Agena D

SPACE TEST PROGRAMME

(USA — DOD)

Role: Space science and engineering
Data: STP Agena D satellite (Astex)
Height: 23 ft 3 in (7.1 m) average without payload
Diameter: 5 ft (1.5 m)
Orbital mass: 1 277 lb (579 kg) without payload
Spacecraft: Multipurpose propulsion module and payload platform carries variable payloads attached to forward end. Previous versions (Agena A and B) required payload to be integrated into main body. Cylindrical body contains 16 000 lb (71.2 kN)-thrust Bell Aerosystems restartable engine and tanks for IRFNA/UDMH propellents, providing final boost to orbit and orbital manoeuvring facilities (see Agena D)
Payload: Unannounced number of tests in Advanced Space-Technology Experiments (Astex) mission included three main experiments. One was roll-up solar array weighing 250 lb (113 kg) which deployed to 32 ft (9.7 m)-long panel generating 1.5 KW. Second comprised 19 different high energy particle detectors and third bounced high frequency signals off ground in ionospheric sounding experiment
Programme: Single announced mission, launched on October 17, 1971 probably introduced STP which is outgrowth from DOD Space Experiments Support Programme (see Aerospace Research Support Programme). Provides facility for science projects and engineering tests not meriting individual launches both for DOD and non-military government experimenters (i.e. Nasa contractors). Uses Agena D satellite, latest variant of Agena self-boosting satellite already used by DOD for more than 300 unidentified missions with only 30 failures. Notable applications included Midas (see IMEWS) and Discoverer. Boosters include Thor and Titan IIIB. Agena also used successfully as second-stage booster on Thor, Atlas and Titan for some 50 launches both civil and military. Two more missions planned for 1978. Prime contractor: Lockheed
Launch vehicle: Thor (Agena)

AEROSPACE RESEARCH SUPPORT PROGRAMME

Role: Space physics and engineering
Data: Orbiting Vehicles (OV)
OV1: Detailed separately
OV2: More complex than OV1 and had only three flights — OV2-1, -3 (both lost at launches in October and December 1965) and -5 (September 1968). All similar 22-24 in (56-61 cm) cuboids with four solar paddles and weighing 375-450 lb (170-204 kg) with multi-experiment physics payloads. Prime contractor: Northrop
OV3: Set of six craft launched by Scout between April 1966 and December 1967 (OV3-3 launch failure) with 6 or 7 space physics experiments each. Spin-stabilised octagonal cylinder structures; first four 29 x 29 in (74 x 74 cm) weighing 152-177 lb (69-80 kg) and two only 21 in (53 cm) high weighing 209 lb and 222 lb (95 kg, 100 kg). Prime contractor: USAF Cambridge Research Laboratories
OV4: Only three units in single launch by Titan 3C in November 1966 used modified Titan 2 tank as OV4-3 to obtain MOL data and carry nine other experiments. Ejected similar subsatellites as OV4-1R (receiver) and -1T (transmitter) for communications experiment (plus OV1-6). Prime contractor: Convair (OV4-3); Raytheon (OV4-1R, -1T)
OV5: See Environmental Research Satellite
Programme: ARSP is part of DOD space research programme managed by US Air Force Office of Aerospace Research to provide inexpensive means of orbiting Air Force Research and technology experiments. Operates jointly with Samso-managed Space Experiments Support Programme which provides facility for civil payloads from other government experimenters. Main series include some subprogrammes, as with Nasa's Explorers. Satellites are launched as multiple or "piggyback" payloads, usually by refurbished Atlas F ballistic missile

ORBITING VEHICLE 1

Role: Space physics and engineering
Data: OV1-19
Length: 2 ft 8 in (0.81 m). With solar arrays: 4 ft 4 in (1.42 m)
Diameter: 2 ft 3 in (0.68 m)
Orbital mass: 273 lb (124 kg)
Spacecraft: Standard platform is cylinder containing 6 cu ft (0.17 m^3) of payload capacity. Hemispherical solar arrays at either end generate 22 W and enclose power and thermal control equipment. Standard solid-fuel propulsion module jettisoned after orbital positioning manoeuvre. Spacecraft spin-stabilised
Payload: Seven experiments to study trapped radiation in the Van Allen belt and five to study hazards to Man of such radiation
Programme: OV1-19 launched in March 1969 with OV1-17, -17A and 18, all manoeuvring into different orbits. 23 OV1s launched between January 1965 and August 1971, mostly as multiple or "piggyback" payloads. Stabilising techniques according to payload needs. Some OV1s carried subsatellites for release in orbit (OV1-21, latest announced launch, released two balloons and a sphere). Others were special designs using OV1 propulsion module. OV1 is most extensively used of five different OV series (see Aerospace Research Support Programme) and includes some civil scientific and engineering payloads. Prime contractor: Convair (General Dynamics)
Launch vehicle: Atlas F

(USA — DOD) ENVIRONMENTAL RESEARCH SATELLITE

Role: Space physics and engineering
Data: OV5-5 (ERS-29)
Orbital mass: 25 lb (10 kg)
Spacecraft: Spin-stabilised, octahedral spacecraft (eight equal triangular faces) measures 12 in (30 cm) along each side, with solar cells covering all faces and generating 4.5 W. Magnetometer boom and 21 in (53 cm) deployable VLF antenna project from points. Design life is 18 months
Payload: 30 experiments provided by VLF plasma wave detector, magnetometer, seven particle detectors and new data and telemetry equipment. Worked with OV5-6 and -9 in collection of solar radiation data
Programme: OV5-5, launched with OV5-6 and -9 in May 1969, was one of several growth configurations in small, inexpensive ERS series. More than 30 launched since September 1962 to collect space physics and engineering data for US Air Force. First ERS form was tetrahedron with 7 in (18 cm) sides weighing 1.47 lb (0.67 kg), increasing to 9 in (23 cm) and 11 in (28 cm) per side and weighing up to 4.5 lb (2 kg). Six successes out of 11 launches were designated Tetrahedral Research Satellites (TRS) 1-6. Continued in octahedral form (ORS) with 8 in (20 cm) sides and 12 lb (5.4 kg) weight for four missions. Subsequent missions incorporated in 1967 into Advanced Research Support Programme as OV5 series 1-9 (seven successful) using 20-30 lb (9-14 kg) versions (OV5-6 and -9 were large tetrahedrons). Three 40 lb (18 kg) versions with 12 in (30 cm) sides also used by Nasa as test and training (TETR) satellites for Apollo ground crews. Series also covers cylindrical Apollo P&F satellites. Latest form is prismatic for Earth survey application. All spacecraft orbited as "piggyback" payloads on a variety of launchers. Prime contractor: TRW
Launch vehicle: Titan IIIC (OV5-5, -6 and -9, Vela 9 and 10 in single launch)

СОЮЗ

Role: Multipurpose assortment
Spacecraft: Cosmos series is blanket designation for a wide variety of civil and military payloads totalling 701 between March 1962 and end of 1974. Few details available but statistical summaries are possible from data analysis
Military Cosmos: Some 574 payloads, comprising almost 82 per cent of Cosmos total, are devoted to military purposes. Of these, 294 represent about four generations of observation satellites, mostly with recoverable capsules and later ones with manoeuvring capability. Remainder include 13 octuple launches, probably for military navigation, plus ferret, communication meteorology and engineering. Also includes 36 launches connected with Fractional Orbit Bombardment System development (now complete)
Civil Cosmos: Remaining 127 payloads are divided almost equally between space sciences/astronomy and engineering. Many would correspond to Nasa's Explorer series and other specialised programmes. Includes manned spacecraft and applications engineering flights plus Earth escape failures (see Soyuz, Meteor, Molniya). Series represents the bulk of Soviet space activities
Launch vehicle: Variable

INTERCOSMOS

Role: Space sciences
Data: Intercosmos 12
Height: Overall approx 19 ft (5.8 m). Main body approx 8 ft (24 m)
Diameter: Main body approx 4 ft 6 in (1.4 m)
Spacecraft: Standard spacecraft structure has elliptical main body with experiment and antenna mast projecting from top and four sets of deployable solar panels fitted round midsection
Payload: First in series to have completely non-Russian experiments including an ionospheric sounder (E. Germany) and a mass-spectrometer calibrator (Romania)
Programme: Series of international satellites using Soviet designed and launched spacecraft with experiments built co-operatively or independently by various Soviet bloc countries. First launch in December 1968 was designated Cosmos 261. Total of 13 Intercosmos launches followed between October 1969 and April 1975 including some sub-groups of associated flights, eg Intercosmos 3, 5, 6 and 10 (magnetospheric studies), Intercosmos 2, 8, 9 and 12 (ionospheric studies) and Intercosmos 1, 4, 7 and 11 (solar radiation and its effects on upper atmosphere). Intercosmos 6 weighed 2,360 lb (1,070 kg) and introduced new series with recoverable payloads. Programme continuing
Launch vehicle: (Intercosmos 12) C-1

PROGNOZ

Role: Space sciences
Data: Prognoz 3
Orbital mass: 1 859 lb (845 kg)
Spacecraft: Sealed main compartment is shallow, round-ended drum surmounted by experiment sensors and containing spacecraft controls. Four deployable solar panels fitted to side of main body, one panel carrying a magnetometer boom on the end while the opposite panel carries a long-wave antenna
Payload: Probably similar to that of Prognoz 2 which had 15 experiments to study solar flares, X-ray emission and effects of solar emissions on the ionosphere. Prognoz 2 also carried two French instruments
Programme: Three spacecraft launched since April 1972. Prognoz 2 launched two months after first and operated in conjunction with it from opposing highly eccentric orbits. Prognoz 3 launched in February 1973 and complemented studies by Lunokhod 2 of lunar magnetic field
Launch vehicle: A-2-e

PROTON

(USSR)

Role: Stellar astronomy
Data: Proton 4
Height: About 12 ft (3.6 m)
Diameter: About 13 ft (4.0 m). Array span: about 32 ft (9.7 m)
Orbital mass: 37 500 lb (17 000 kg)
Spacecraft: Main body is short cylinder or annulus from which projects a truncated cone of thermal blanketing. On top, a mounting ring carries four triangular, deployable solar panels. A convex base also projects from within the body shell. Design life is a few months
Payload: Single experiment is a massive cosmic ray detector weighing 27 500 lb (12 500 kg) capable of registering cosmic rays up to 10^{15} electron volts and the nuclear interaction of very high energy cosmic particles. Continued work of preceeding Protons
Programme: The first three protons, launched in July and November 1965 and July 1966, were similar to Proton 4 (launched in November 1968) but with a low convex upper projection similar to the base and weighing 26 896 lb (12 200 kg). Proton 1 was first publicised use of D launcher. Programme appears concluded but no clear indication has been made of this
Launch vehicle: D-1

APPLICATIONS TECHNOLOGY SATELLITE

Role: Applications technology
Data: ATS 5
Height: 6 ft (1.8 m)
Diameter: 5 ft (1.5 m)
Orbital mass: 780 lb (354 kg). Launch weight: 954 lb (433 kg)
Spacecraft: Geostationary spacecraft has cylindrical main body covered in solar cells generating 185 W. Internally mounted 6,250 lb (27.8 kN)-thrust solid-fuel apogee motor jettisons in orbit. Gravity gradient stabilised
Payload: Total of 13 experiments is largest number and heaviest payload carried in series. Includes four 124 ft (38 m)-long deployable gravity gradient booms forming an X with variable angles, plus two 45 ft (14 m)-long damper booms projecting diagonally through spacecraft. Also carries TV, C-band and L-band propogation experiments, six different reaction control systems, including an ion engine, plus three particle detectors
Programme: First generation ATS is outgrowth of Advanced Syncom project and comprised five similar spacecraft launched between August 1966 and August 1969. Series equipped to test 20 different elements of communication, navigation, meteorology and spacecraft control technology. ATS 1 placed in medium orbit, remainder intended for geostationary locations. ATS 2 and 4 failed to achieve geostationary orbit but provided some data (ATS 4 decayed after two months). ATS 3 and 4 were spin-stabilised and needed only 56 in (1.4 m)-high bodies. Remainder had gravity-gradient stabilisation which required larger bodies for power increase. ATS 5 entered inadvertant spin, preventing boom deployment and operation of four experiments. Prime contractor: Hughes.
Launch vehicle: (ATS 5) Atlas Centaur

APPLICATIONS TECHNOLOGY SATELLITE 6

(USA — Nasa)

Role: Communications technology
Data: ATS 6
Height: 26 ft (8.0 m)
Diameter: (Array span) 52 ft (16.0 m)
Launch weight: 3 090 lb (1 402 kg)
Spacecraft: Geostationary satellite has rectangular Earth Viewing Module (EVM) as main spacecraft body containing experiments controls and three-axis stabilisation system. An open truss separates it from a 30 ft (9 m) diameter deployable dish antenna surmounted by a hub carrying an Environmental Measuring Experiments package (EME) and two arms bearing a pair of 10 ft (3.0 m)-wide semicylindrical solar arrays. Incorporates several techniques innovative to comsats including passive thermal control. Design life is two years
Payload: Equipped for more than 20 experiments, mostly concerning direct broadcasting but also including high-resolution radiometer for meteorology, sensors for analysis of space environment at synchronous altitude and general technology experiments
Programme: Single spacecraft launched in May 1974 is second-generation ATS. Initially used in US remote medical diagnosis experiment and was moved in mid-1975 to cover India for educational TV experiment. Designed for use with very small ground antennae serving small communities. Second spacecraft cancelled through lack of Nasa funds. Prime contractor: Fairchild Industries
Launch vehicle: Titan IIIC

EARTH RESOURCES TECHNOLOGY SATELLITE (USA — Nasa)

Role: Earth survey technology
Data: Erts 1 (Landsat 1)
Height: 10 ft (3.05 m) excluding arrays
Diameter: 5 ft (1.52 m). Array span: 11 ft (3.53 m)
Launch weight: 1 965 lb (891 kg)
Spacecraft: Butterfly-shaped craft is based on Nimbus design and comprises circular lower section housing payload and hexagonal upper section containing three-axis stabilisation and other control subsystems. An open truss structure separates the two sections. Two 4 ft x 8 ft (1.2 m x 2.4 m) arrays generate 200 W. Design life is two years
Payload: Two main sensors provide images of the Earth's surface in different spectra. Return Beam Vidicon comprises three cameras operating in green, red and near infra-red with 330 ft (100 m) resolution. Multi-Spectral Scanner images in green, red and two infra-red bands with resolution of 230 ft (70m). Both cover the same 115-mile (185 km)-wide track and false-colour pictures are reconstructed from monochrome transmissions
Programme: Two spacecraft launched into polar orbit in July 1972 and January 1975. Success of Erts 1, in spite of loss of RBV soon after launch, delayed launch of second craft. Data received directly by terminals in 37 countries. The value of this pioneering attempt at remote sensing from space exceeded all expectations, and developments will be more in area of analysing data received than on satellite technology itself. Third Erts is planned for launch in 1977. Prime contractor: General Electric
Launch vehicle: Delta 1914

(USA — DOD)

VELVA

Role: Nuclear test detection
Data: Vela 11/12
Diameter: 4 ft 2 in (1.27 m)
Orbital mass: 571 lb (259 kg)
Spacecraft: Identical spacecraft are polyhedrons each with 26 facets, 24 of which carry solar arrays together generating 120 W. Central cylinder contains 7 600 lb (33.8 kN)-thrust solid-fuel apogee motor. Three-axis stabilisation provided by reaction wheel, rate gyro and nitrogen gas jets, also used for orbital positioning
Payload: Dual-purpose payload is primarily intended to detect nuclear detonations on the Earth and up to 100 million miles (160 million km) from Earth. Also provides data on solar radiation for manned space activities. Total of 162.5 lb (74 kg) of instruments includes Earthward-looking pod within spacecraft body and 28 detectors of 19 different types fitted on corners, including X-ray, gamma-ray, neutron, optical, electromagnetic pulse and energetic particle detectors
Programme: Last pair in set of six Advanced Velas all launched in three pairs between April 1967 and April 1970, some with ERS "piggyback" payloads. Series was preceded by six smaller first-generation Velas, also launched in pairs between October 1963 and July 1965. After insertion into transfer orbit, each pair fired apogee motors to position themselves in circular orbits on opposite sides of the Earth outside Earth's radiation belts. Velas 11 and 12 orbits were at 69 000 miles (111 000 km) altitude. Prime contractor: TRW
Launch vehicle: Titan IIIC (Vela 11 and 12 were first operational launch)

MISSILE EARLY WARNING SYSTEM

(USA — DOD)

Role: Defence early warning
Data: IMEWS
Height: (unconfirmed) 9 ft 10 in (3.0 m)
Diameter: (unconfirmed) 9 ft 10 in (3.0 m). Array span: 23 ft (7.0 m)
Launch weight: (unconfirmed) 1 800 lb (820 kg)
Spacecraft: Reported to be cylinder with four deployable solar arrays arranged in cruciform. Three-axis stabilised in geostationary orbit. Half launch weight is solid fuel for apogee motor
Payload: Equipped with infra-red telescope capable of detecting heat of missile launch. Also reported to carry TV camera able to transmit pictures of detected missiles
Programme: Integrated Missile Early Warning Satellite is third generation system with four spacecraft launched to date between November 1970 and June 1973. IMEWS 1 failed to achieve proper orbit. Believed to be first US operational geostationary system replacing experimental geostationary system referred to as BMEWS (Ballistic Missile Early Warning System) for which four spacecraft were launched between August 1968 and September 1970 and two others in December 1971 and March 1973. Preceded by Midas (Missile Defence Alarm System) series of instrumented Agena stages launched into 2 000-mile (3 200 km) orbits. Launch announced of Midas 1, 2, 3 and 4 between February 1960 and October 1961. USAF classified launches also believed to include at least six more (April 1962, May and July 1963. August and October 1966 and June 1967). Prime contractor: TRW (IMEWS)
Launch vehicle: Titan IIIC

TRANSIT

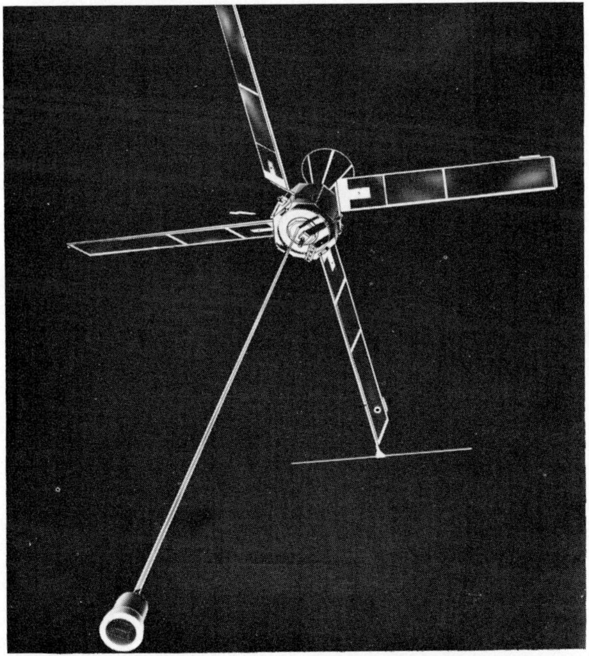

Role: Navigation
Data: Transit 5A
Height: 12 in (0.3 m)
Diameter: 18 in (0.46 m)
Orbital mass: 135 lb (61 kg)
Spacecraft: Octagonal glass fibre body with four solar arrays deploying from sides. Gravity gradient boom with damping spring and end mass deploys from small cylindrical container on top. Conical antenna fitted below spacecraft
Payload: Two ultra-stable oscillators operating on 150 MHz and 400 MHz plus digital clock. Provides dopplar signals for relative position fixing by receiving station, with constant ephemeris updates
Programme: Transit development programme comprised seven launches between September 1959 and November 1962 including two launch failures (Transit 1A and 3A). Transit 4 back-up (T4B) also launched. Transit 5A, which failed on second day after launch in December 1962, was first of two prototypes for basic four-satellite operational system and completely new design from predecessors. Programme subsequently classified until 1967 when system was made available for civil use. At least ten unidentified Transit-related and navigation-technology launches between September 1963 and September 1967. Launches in March 1968, August 1970, October 1973 designated Navsat (Naval Navigation Satellite) 18, 19 and 20. Six satellites in service at end of 1974. Prime contractor: John Hopkins University
Launch vehicle: Scout

GLOBAL POSITIONING SYSTEM

Role: Navigation
Data: Navstar
Array Span: 16 ft 11 in (5.1 m)
Orbital mass: 800 lb (363 kg). Launch weight 1 400 lb (635 kg)
Spacecraft: Spin-stabilised spacecraft has cylindrical main body covered in solar cells plus pair of deployable solar arrays on despun mast projecting from top. Design life is five years
Payload: Atomic clock and two transmitters, one operating at 1 200 MHz and the other between 1 500 and 1 600 MHz
Programme: Phase one of second-generation Navsat network involves five Navstar (Navigation System using Time And Ranging) satellites and one Navigation Technology Satellite (NTS-2) launched into two 10 000 miles (16 000 km) circular planes in 1977-78. To be followed in 1981-82 by phase two, adding five more satellites and a third orbital plane. Phase three will complete GPS system in 1984 by bringing total to 24 satellites in three planes. Position fixing achieved by receiver measuring signal delay and is based on technology from several classified USAF satellites launched between 1964 and 1968 plus Timation 1, 2 and 3 (NTS-1) (May 1967, September 1969 and July 1974) and US Navy's Triad launched in September 1972 to test automatic decay compensating device. GPS will also be available for civil use. Prime contractor: Rockwell
Launch vehicle: Atlas F

(USA — Comsat)

COMSAT DOMESTIC

Role: Domestic satellite communications
Data: Comstar
Height: 9 ft 3 in (2.8 m). With antennae: 19 ft 8 in (6.0 m)
Diameter: 7 ft 9 in (2.3 m)
Orbital mass: 1 725 lb (782 kg). Launch weight: 3 283 lb (1 489 kg)
Spacecraft: Similar to Intelsat 4
Payload: Communications repeater of 24 transponders with six kept as spare and remainder providing up to 28 000 voice channels. Uses similar antenna technique to Intelsat 4A in which clusters of feeder horns serve single pair of spot-beam antennae. This eliminates sidelobe interferance and doubles capacity by allowing use of same frequency for east and west transmissions
Programme: First of two spacecraft is to be launched in 1976 to provide commercial US domestic comsat network. System to be operated by Comsat General and facilities leased by AT&T. Prime contractor: Hughes
Launch vehicle: Atlas-Centaur

(USA — RCA)

RCA SATCOM

Role: Domestic satellite communications
Data: RCA Satcom Phase II
Height: 5 ft 4 in (1.6 m)
Width: 4 ft 1 in (1.2 m) square
Orbital mass: 1 000 lb (454 kg). Launch weight 2 000 lb (908 kg)
Spacecraft: Geostationary satellite is rectangular box with two deployable solar panels. Three-axis stabilisation. Design life is eight years
Payload: Communications repeater of 24 transponders operating in the 4 GHz and 6 GHz wavebands. Each transponder can carry either one TV channel, 1 000 FDM/FM voice channels or 600 single channel per carrier voice circuits
Programme: US domestic comsat network for RCA Globcom will use three geostationary satellites launched by a specially uprated Delta model. First launch scheduled for late 1975 network will replace present Phase I where RCA leases space aboard Canada's Anik satellites. Prime contractor: RCA
Launch vehicle: Delta 3914

WESTAR

(USA—Western Union Telegraph Co)

Role: Domestic communications
Data: Westar
Height: 5 ft 1 in (1.6 m). Overall: 11 ft 8 in (3.5 m)
Diameter: 6 ft (1.8 m)
Orbital mass: 625 lb (283 kg). Launch weight: 1 265 lb (574 kg)
Spacecraft: Geostationary spin-stabilised satellite based on Anik design. Body-mounted solar array and hydrazine-jet attitude control. Design life is seven years

Payload: Communications repeater of 12 transponders providing a maximum 14 000 or average of 6 000 telephone channels or up to 12 colour TV channels. Single electrically despun dish antenna covers US mainland plus Hawaii
Programme: First independent domestic comsat network comprises two spacecraft launched in April and October 1974 and operated by Western Union Telegraph Co. Prime contractor: Hughes
Launch vehicle: Delta 2914 "Straight Eight"

OSCAR

Role: Amateur radio communications
Data: Oscar 7
Height: 17 in (43 cm)
Diameter: 16.5 in (42 cm)
Orbital mass: 65 lb (29.5 kg)
Spacecraft: Second in new generation of Oscar series is eight-sided cylinder spin-stabilised and covered in solar cells. Design life is three years
Payload: Communications repeater has two transponders operating on 29.5 MHz and 145.95 MHz downlink frequencies plus two radio beacons operating on 435.1 MHz and 2 304.1 MHz.
Programme: Latest in series of Orbital Satellite Carrying Amateur Radio (Oscar) produced and operated by Radio Amateur Satellite Corporation (Amsat) for unrestricted use by radio "hams" throughout the world. Oscar 7 is third to be solar powered, previous models having battery power for two month's operation. Seven satellites launched mostly in multiple payloads between December 1961 and November 1974. First five launched by USAF, remainder by Nasa. Prime contractor: Amsat
Launch vehicle: (Oscar 7) Two-stage Delta

TELESAT

(Canada)

Role: Domestic communications
Data: Anik 1
Height: 5 ft 1 in (1.6 m). Overall: 11 ft 6 in (3.5 m)
Diameter: 6 ft 3 in (1.9 m)
Orbital mass: 650 lb (295 kg). Launch weight 1 250 lb (567 kg)
Spacecraft: Geostationary spacecraft, spin-stabilised at 60 rpm. Design based on Intelsat 4. Body-mounted solar array generates 250 W. Design life is seven years

Payload: Communications repeater of 12 transponders operating at 33 dBW Effective Radiated Power through a single 60 in (1.5 m) dish antenna. Provides 690 telephone circuits or a colour TV channel per transponder

Programme: World's first domestic comsat network comprises two spacecraft launched in November 1972 and April 1973 to provide Canada with link to far North. Back-up spacecraft also launched in March 1975. Prime contractor: Hughes
Launch vehicle: Delta 1914 "Straight Eight"

COMMUNICATIONS TECHNOLOGY SATELLITE

Role: CTS/CAS-C
Height: 3 ft 10 in (1.2 m)
Diameter: 5 ft 11 in x 4 ft 7 in (1.8 m x 1.4 m). Array span: 51 ft 6 in (16.0 m)
Orbital mass: 792 lb (352 kg)
Spacecraft: Three-axis stabilised geostationary satellite with two unfurlable solar arrays generating 1.2 KW plus body-mounted arrays generating 77 W. Design life is two years
Payload: Very high power transponder (200 W) operating through pair of 2.5° gimballed antennae in the 11 GHz and 14 GHz frequency ranges to test use of miniature (3 ft) Earth terminals. Also carries an SHF beacon and is the first spacecraft to test SHF frequencies. Carries US-supplied power tube and solar arrays in return for Nasa launch. Also carries some Esro equipment for testing.
Programme: Single experimental comsat was to be launched early 1976. Also given Nasa designation Co-operative Applications Satellite (CAS) -C. Prime contractor: Communications Research Centre
Launch vehicle: Delta 2914

(ESA)

ORBITAL TEST SATELLITE

Role: Communications technology
Data: OTS
Height: 3 ft 10 in (1.18 m) Overall 7 ft 6 in (2.3 m)
Diameter: (Maximum) 7 ft 1 in (2.18 m). Array span: 30 ft 4 in (9.26 m)
Orbital mass: 979 lb (444 kg). Launch weight: 1 909 lb (866 kg)
Spacecraft: Three-axis stabilised geostationary spacecraft with two solar arrays generating 700 W. Six-sided main body (two long sides and four short sides) contains solid-fuel apogee motor. Modular design can carry variable payload. Design life is three years
Payload: Multi-experiment communications payload uses six 20 W transponders including one spare. "Module A" provides for telephone and TV tests in the 11 GHz and 14 GHz frequency ranges using two 40 MHz bandwidth channels to provide European/N. Africa regional coverage through a pair of 7.5° x 4.25° "Eurobeam A" dish antennae, and two 120 MHz bandwidth channels using the same size reception antenna but transmitting through a 2.5° steerable spotbeam antenna. "Module B" uses a 5 MHz channel operating through a pair of 5° x 3.5° "Eurobeam B" antennae to test use of small ground terminals of 10 ft (3 m) diameter
Programme: Single experimental satellite originated by UK and adopted by Europe as precurser to European regional comsat planned for early 1980s. Scheduled for launch in early 1977. Prime contractor: HSD (for Mesh consortium)
Launch vehicle: Delta 3914

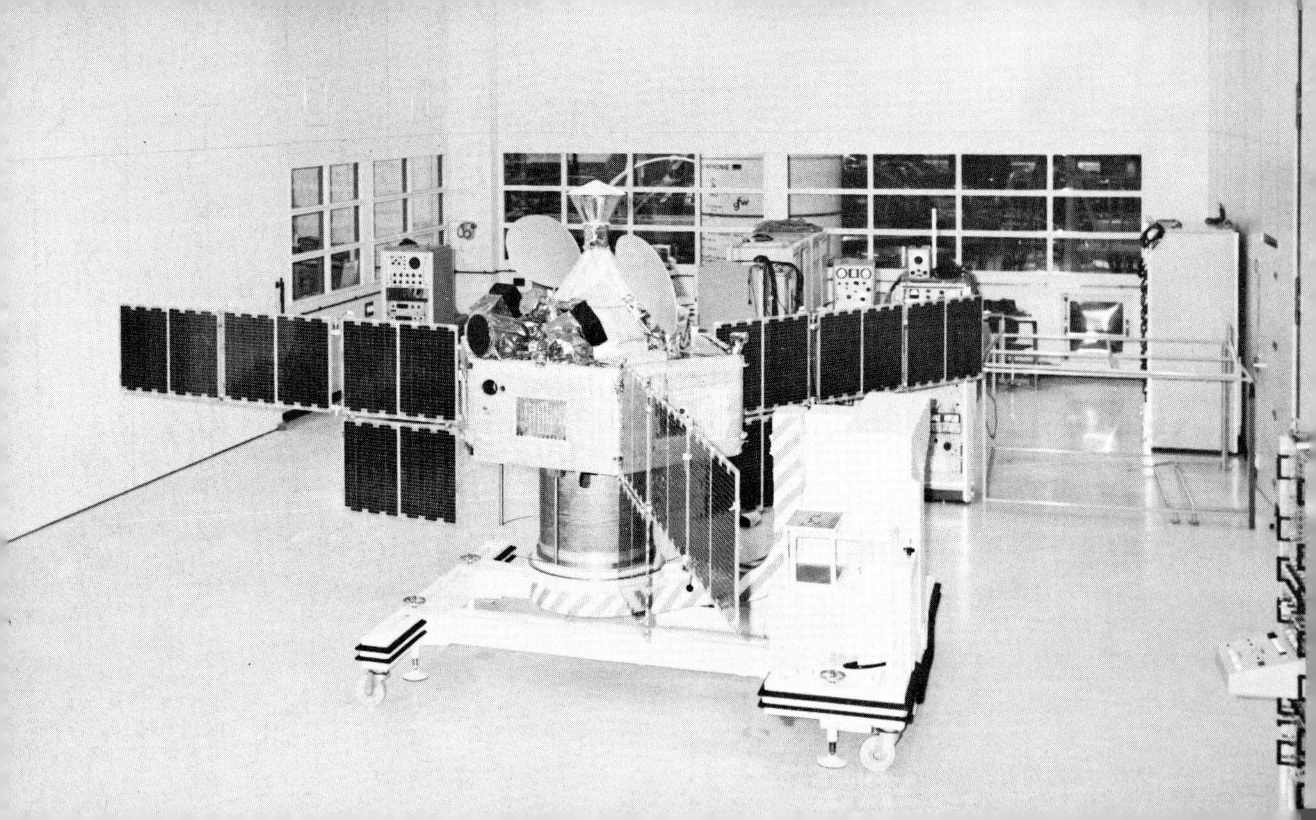

SYMPHONIE

(W. Germany/France)

Role: Communications technology
Data: Symphonie 1
Height: 1 ft 11 in (0.58 m). Overall 6 ft 6 in (2.0 m)
Diameter: 5 ft 6 in (1.7 m). Array span: 22 ft 5 in (6.86 m)
Orbital mass: 521 lb (237 kg). Launch weight 886 lb (402 kg)
Spacecraft: Geostationary satellite, three-axis stabilised. Three deployable solar arrays attached to six-sided main body generate 300 W. An 88 lb (392 N)-thrust liquid propellant apogee motor (nitrogen tetroxide/aerozine 50) is fitted to base. Station-keeping provided by seven bipropellant thrusters, attitude control by eight cold-gas thrusters. Design life is five years
Payload: Communications repeater of two 13 W transponders providing 600 telephone channels or two TV channels. Also carries experimental Time-Division Multiple Access system. Twin 13° x 88° spot-beam transmission antennae illuminate north and south Atlantic seaboards. Twin horn antennae for US reception
Programme: First of two identical spacecraft launched in December 1974 over South Atlantic. Second was launched in August 1975. This first independent international comsat evades conflict with Intelsat charter by retaining experimental status. Prime contractor: Aerospatiale (for Cifas consortium).
Launch vehicle: Delta 2914

SIRIO

(Italy)

Role: Communications technology
Data: Sirio
Height: 2 ft 11 in (0.95 m). Overall: 6 ft 6 in (1.99 m)
Diameter: 4 ft 8 in (1.43 m)
Orbital mass: 485 lb (220.5 kg)
Spacecraft: Geostationary satellite spin-stabilised at 90 rpm. Body-mounted solar array generates 118 W. Hydrazine-jet attitude control and solid-fuel apogee motor. Design life is two years
Payload: Two 9 W transponders operating in 12 GHz and 18 GHz frequency ranges through single mechanically despun SHF antenna. Designed for general and multiple access communications activities
Programme: Single experimental comsat was to be launched in mid-1975. Prime contractor: Compagnia Industriale Aerospaziale
Launch vehicle: Delta

EXPERIMENTAL COMMUNICATIONS SATELLITE

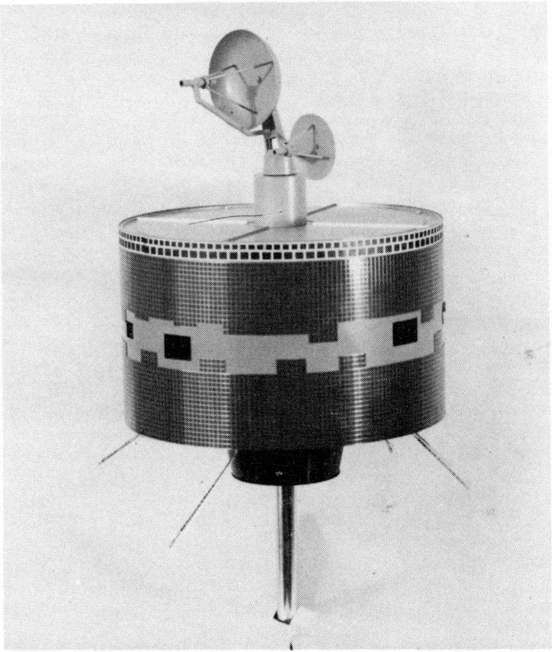

Role: Communications technology
Data: ECS
Height: Main body 47 in (1.2 m)
Diameter: 55 in (1.4 m)
Orbital mass: 287 lb (130 kg)
Spacecraft: Drum-shaped spacecraft is spin-stabilised and has circumference covered in solar cells.
Payload: Active communications repeater for experimental communications at both 6/4 gHz and 20/30 gHz. Antenna assembly is electrically despun and comprises two spot-beam dish aerials
Programme: ECS is planned for launch into geostationary orbit in 1977. Launch is to be preceded by structurally similar Engineering Test Satellite (ETS)-2 in 1976 which will be first geostationary attempt by Japanese national launch vehicle. Prime contractor: Mitsubishi
Launch vehicle: N

COMMUNICATIONS SATELLITE (Japan — Nasda)

Role: Communications technology
Data: Medium-Capacity Satellite for Experimental Purposes (CS)
Height: Overall 12 ft 6 in (3.8 m)
Diameter: 7 ft 3 in (2.2 m)
Orbital mass: 750 lb (340 kg)
Spacecraft: Cylindrical spacecraft is spin-stabilised and has body-mounted solar array. Single despun directional aerial projects from top
Payload: Active communications repeater providing operational tests of voice and data links plus some subsystem engineering
Programme: Single spacecraft is scheduled for launch into geostationary orbit in November 1977 to provide Japan with domestic communications facility. Prime contractor: Mitsubishi
Launch vehicle: Delta 2914

(Japan—Nasda)

BROADCASTING SATELLITE

Role: Communications technology
Data: Medium-scale Broadcasting Satellite for Experimental purposes (BSE)
Height: 9 ft 10 in (3.0 m)
Diameter: 7 ft 3 in x 5 ft 3 in (2.2 m x 1.6 m)
Orbital mass: 772 lb (350 kg)
Spacecraft: Rectangular spacecraft is three-axis stabilised and has two deployable solar arrays fitted to sides. Single large dish aerial is fitted on top. Design life is three years
Payload: Communications repeater with three transponders, two of which are to be used for transmission of two colour television channels and third kept as spare
Programme: Single spacecraft is scheduled for launch in 1978 to provide experimental domestic broadcasting service. Prime contractor: Toshiba
Launch vehicle: Delta 2914

MOLNIYA 1

(USSR — Intersputnik)

Role: International communications
Data: Molniya 1
Height: 11 ft 3 in (3.4 m)
Diameter: 5 ft 2 in (1.55 m)
Launch weight: About 2 200 lb (1 000 kg)
Spacecraft: Three-axis stabilised spacecraft consists of cylindrical main body surmounted by a conical housing for Earth and Sun sensors and with six deployable arrays of solar panels fitted round lower end which contains liquid-fuel restartable orbit correction engine of 441 lb (1.96 kN) thrust. Operational life estimated at five-six years
Payload: Communications repeater and space relay system of three transponders, including two spare. Provides transmission of one TV channel at 40 W or an unspecified number of telephone channels at 14 W through one of duplicated pair of 22°, 3 ft (0.9 m) dish antennae set 180° apart. The spare antenna can be brought into use by rotating spacecraft. Narrow beam transmission used to reduce Earth terminal size
Programme: 28 Molniya 1s launched between April 1965 and end of 1974 (not including failures — Cosmos 259?). Satellites are in highly inclined, 12 h elliptical orbits with apogee over the USSR, remaining in view for 11 h, with consecutive craft overlapping to provide uninterrupted service. Programme apparently continuing. Lack of equatorial launch facilities plus the need to cover high latitudes have dictated this technique rather than geostationary network. Intersputnik is a nine-nation socialist consortium established in 1971
Launch vehicle: A-2-e

(USSR — Intersputnik) MOLNIYA 2

Role: International communications
Data: Molniya 2
Height: 13 ft 6 in (4.1 m)
Diameter: 5ft (1.5 m)
Launch weight: About 2 750 lb (1 250 kg)
Spacecraft: Similar to Molniya 1 but slightly enlarged and with new antennae
Payload: Similar to Molniya 1 but operating on frequencies in the 4 GHz and 6 GHz wavebands, as used by Intelsat
Programme: Second generation comsat. 11 Molniya 2s launched between November 1971 and end of 1974 and still ongoing. Satellites appear to operate as a pair interspersed with a pair of Molniya 1s, the four set 90° apart with replacement craft being sent up as and when necessary. The highly eliptical orbit ensures that one of each is always in view of the USSR at any one time. This replaces a three-satellite system used with Molniya 1. The first of a third generation comsat, Molniya 3, was launched in November 1974. Also, the first Soviet geostationary satellite, a comsat designated Molniya 1S, was launched in July 1974. Network of three geostationary comsats planned for launch over Indian Ocean in 1975/76 comprising Statsionar T (USSR TV only), Statsionar 2 (Europe and Western USSR telecommunications) and Statsionar 3 (USSR telecommunications only)
Launch vehicle: (Molniya 2) A-2-e

(ESA) MARITIME ORBITAL TEST SATELLITE

Role: Maritime communications technology
Data: Marots
Height: 3 ft 10 in (1.18 m). Overall 6 ft 4 in (1.95 m)
Diameter: (maximum 7 ft 1 in (2.18 m). Array span: 27 ft 11 in (8.6 m)
Orbital mass: 1 027 lb (466 kg). Launch weight 1 958 lb (888 kg)
Spacecraft: Three-axis stabilised geostationary satellite with two deployable solar arrays plus body-mounted arrays generating 600 W. Uses same modular spacecraft unit as OTS comprising six-sided frame of two long and four short sides containing control subsystems and solid-fuel apogee motor. Design life is at least three years
Payload: Communications repeater using two transponders operating through single 78 in (2.0 m), 16° dish antenna. One transponder provides high power (80 W) shore-ship link and the other provides low power (4.5 W) ship-shore link, exploiting large land terminals. Shore-satellite segment operates on frequencies in 11 GHz and 14 GHz bandwidths (X-band) while the ship-satellite segment operates on L-Band frequencies between 1.54 GHz and 1.64 GHz. Facilities include voice, data, telex and distress signals
Programme: Single experimental maritime comsat due for launch in 1976 to obtain information for possible international operational system in 1980s. Prime contractor: HSD (spacecraft); Marconi (payload)
Launch vehicle: Delta 3914

MARISAT

(USA — Comsat)

Role: Maritime communications
Data: Marisat
Height: 5 ft 3 in (1.6 m). Overall 12 ft 4 in (3.76 m)
Diameter: 7 ft 1 in (2.16 m)
Orbital mass: 720 lb (327 kg). Launch weight 1 445 lb (655 kg)
Spacecraft: Geostationary spacecraft has drum-shaped main body spin-stabilised and covered in solar cells generating 330 W. "Anik"-type apogee motor assisted in orbit circularisation by two 5 lb (22 N) yaw thrusters from hydrazine attitude control system. Design life is five years
Payload: Communications repeater system comprises three transponders operating in UHF (240-400 MHz), L-band (1.4-1.6 GHz) and C-band (4 and 6 GHz). UHF provides one voice and 44 teletype channels or 14 voice channels at up to 28 dBW effective radiated power L- and C-bands provide 14 and 22 teletype channels at ERPs of up to 29.5 dBW (C-band) and 18.8 dBW (L-band). Despun antenna "farm" weighs 26 lb (12 kg) and comprises three large helical UHF antennae, four small helical L-band antennae and pair of C-band horn antennae
Programme: First maritime comsat network comprises two satellites to cover Atlantic and Pacific Oceans. First launch was due late 1975. US Navy has leased UHF transponder for two years as stop-gap until Fltsatcom enters service. C-band provides satellite-shore links and L-band for satellite-ship links for commercial shipping. Prime contractor: Hughes
Launch vehicle: Delta 2914

INTELSAT 3

Role: International communications
Data: Intelsat 3
Height: 3 ft 5 in (1.04 m). Overall 6 ft 7 in (2.0 m)
Diameter: 4 ft 8 in (1.4 m)
Orbital mass: 334 lb (151 kg). Launch weight 647 lb (293 kg)
Spacecraft: Geostationary satellite, spin-stabilised at 90 rpm. Body-mounted solar array generates 470 W. Hydrazine attitude control system, 3 140 lb (14 kN)-thrust apogee motor. Design life is five years
Payload: Communications repeater of four transponders provides 1,200 telephone circuits or four TV channels. Single mechanically-despun horn antenna provides global coverage at up to 22 dBW Effective Radiated Power
Programme: Third generation commerical comsat replacing single Intelsat 1 (Earlybird) and first global network of three intelsat 2s. Eight identical Intelsat 3s launched between September 1968 and July 1970 included three launch or orbital positioning failures (F.1, F.5 and F.8). Two satellites still operational at end 1974. Prime contractor: TRW
Launch vehicle: LTTA Delta N

INTELSAT 4

(Intelsat)

Role: International communication
Data: Intelsat 4
Height: 9 ft 3 in (2.8 m). Overall 17 ft 4 in (5.27 m)
Diameter: 7 ft 10 in (2.4 m)
Orbital mass: 1 610 lb (730 kg). Launch weight: 3 120 lb (1 415 kg)
Spacecraft: Geostationary satellite, spin-stabilised at 60 rpm. Body-mounted solar array gives 470 W. Hydrazine-jet attitude control. 12 400 lb (51.2 kN)-thrust apogee motor. Design life is seven years
Payload: Communications repeater of 12 transponders providing 9 000 telephone circuits or 12 colour TV channels (6 000 circuits in practice). Introduced use of spot-beam antennae where two steerable 50 in (1.27 m)-diameter 4.5° antennae illuminate selected regions at 33.7 dBW Effective Radiated Power output. Also, duplicate pairs of horn antennae give global coverage at 22 dBW ERP. Complete antennae platform despun electrically to maintain constant pointing
Programme: Fourth generation commercial comsat for global network (see Intelsat 3). Eight identical satellites launched between January 1971 and mid 1975 of which seven are still in service (one lost during launch). Prime contractor: Hughes
Launch vehicle: Atlas Centaur

INTELSAT 4A

Role: International Communications (civil)
Data: Intelsat 4A
Height: 9 ft 3 in (2.8 m). Overall 22 ft 3 in (7.1 m)
Diameter: 7 ft 9 in (2.3 m)
Orbital mass: 1 760 lb (798 kg). Launch weight: 3 280 lb (1 489 kg)
Spacecraft: Similar to Intelsat 4 but with additional solar cells on top of drum-shaped body
Payload: Communications repeater system with 20 transponders providing 11 000 telephone circuits or 20 colour TV channels. Based on Intelsat 4, the increased capacity comes from a new antennae system which permits use of same frequency for East and West transmissions. Twin 53-in (1.34 cm) spot-beam transmit dish antennae are fed by clusters of 12 horns which eliminate mutual interference. Single reception antenna is similarly treated
Programme: The first of three spacecraft (possibly plus back-up model) was launched in September 1975, the network to provide a stopgap service until an Intelsat 5 can be introduced. Prime contractor: Hughes
Launch vehicle: Atlas Centaur

SKYNET

(UK — MOD)

Role: International communications
Data: Skynet II
Height: (overall) 6 ft 10 in (2.1 m)
Diameter: 6 ft 3 in (1.9 m)
Orbital mass: 517 lb (235 kg). Launch weight: 959 lb (435 kg)
Spacecraft: Geostationary satellite spin-stabilised at 90 rpm. Hydrazine attitude-control system. Body-mounted solar array generates 200 W. Design life is three years
Payload: Duplicated communications repeaters each with single transponder split into two channels to enable communication with large Earth terminals and small (3½ ft) ship-borne or transportable terminals. Mechanically despun horn antenna provides Earth-coverage on 7 GHz and 8 GHz Wavebands (X-band). TT&C uses S-Band for compatibility with US DSCS system. Effective radiated power more than three times that of Skynet 1
Programme: Second generation UK military comsat and first comsat to be developed outside USA and USSR. Two spacecraft built and launched in January and December 1974 to link UK with far East bases and ships. First satellite lost in launch failure but second now in service. Prime contractor: Marconi
Launch vehicle: Delta 2314

NATO 2

Role: International communications
Data: Nato 2B
Height: (overall) 5 ft 3 in (1.6 m)
Diameter: 4 ft 6 in (1.37 m)
Orbital mass: 285 lb (129 kg)
Spacecraft: Geostationary satellite spin-stabilised at 90 rpm. Body-mounted solar array. Design life is five years
Payload: Communications repeater of single transponder split into two channels for operation with large and small antennae. Voice and telex links provided by single mechanically despun Earth-coverage horn antenna
Programme: Nato military comsat linking Nato capitals and units from Turkey to eastern US seaboard. First generation facility in spite of designation. Two-satellite programme — Nato 2A launched in March 1970 and ceased operating in 1972; Nato 2B launched in February 1971 and still operating. Prime contractor: Philco Ford
Launch vehicle: T-A Delta

NATO 3

(Nato)

Role: International communications
Data: Nato-3
Height: 7 ft 4 in (2.24 m). Overall: 10 ft 2 in (3.1 m)
Diameter: 7 ft 2 in (2.18 m)
Orbital mass: 675 lb (306 kg)
Spacecraft: Spin-stabilised geostationary satellite with body-mounted solar array and solid-fuel apogee motor. Design life is seven years
Payload: Communications repeater with three channels for voice, data and telex links. Mechanically despun antennae provide broad-beam coverage of Nato countries and spot-beam coverage of western Europe alone
Programme: Second generation Nato military comsat system involves two spacecraft to be launched in late 1975 and early 1976. Back-up model may also be launched later. Prime contractor: Philco Ford
Launch vehicle: Delta

DEFENCE SATELLITE COMMUNICATIONS SYSTEM 2

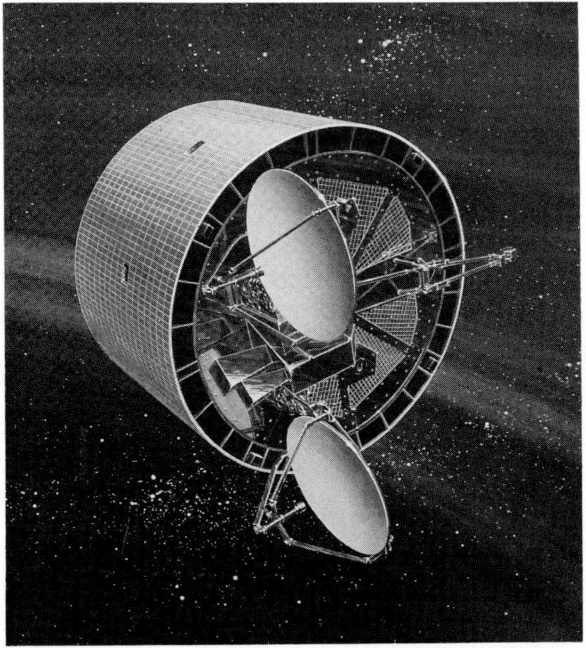

Role: International communications
Data: DSCS 2
Height: 6 ft (1.74 m). Overall: 13 ft (3.95 m)
Diameter: 9 ft (2.75 m)
Orbital mass: 1 100 lb (499 kg)
Spacecraft: Geostationary spin-stabilised satellite with body-mounted solar array generating 520 W. Design life is five years
Payload: Communications repeater of two transponders providing up to 1 300 voice channels on X-band (14 GHz and 16 GHz) frequencies. Operates through pair of Earth-coverage 18° horn antennae at up to 28 dBW Effective Radiated Power and through pair of steerable 44 in (1.1 m) dish antennae producing 2.5° spot beams at up to 40 dBW ERP
Programme: US Department of Defence strategic communications network replacing DSCS 1 system comprises six spacecraft to provide basic three-satellite global coverage plus spares. Satellites can be moved to new locations as required. All launched as pairs in November 1971, December 1973, May 1975. Third pair lost in launch failure and only one of second pair is still operating. Six more ordered with one spot beam modified for wider coverage. First pair due for launch in 1977. Prime contractor: TRW
Launch vehicle: Tital IIIC

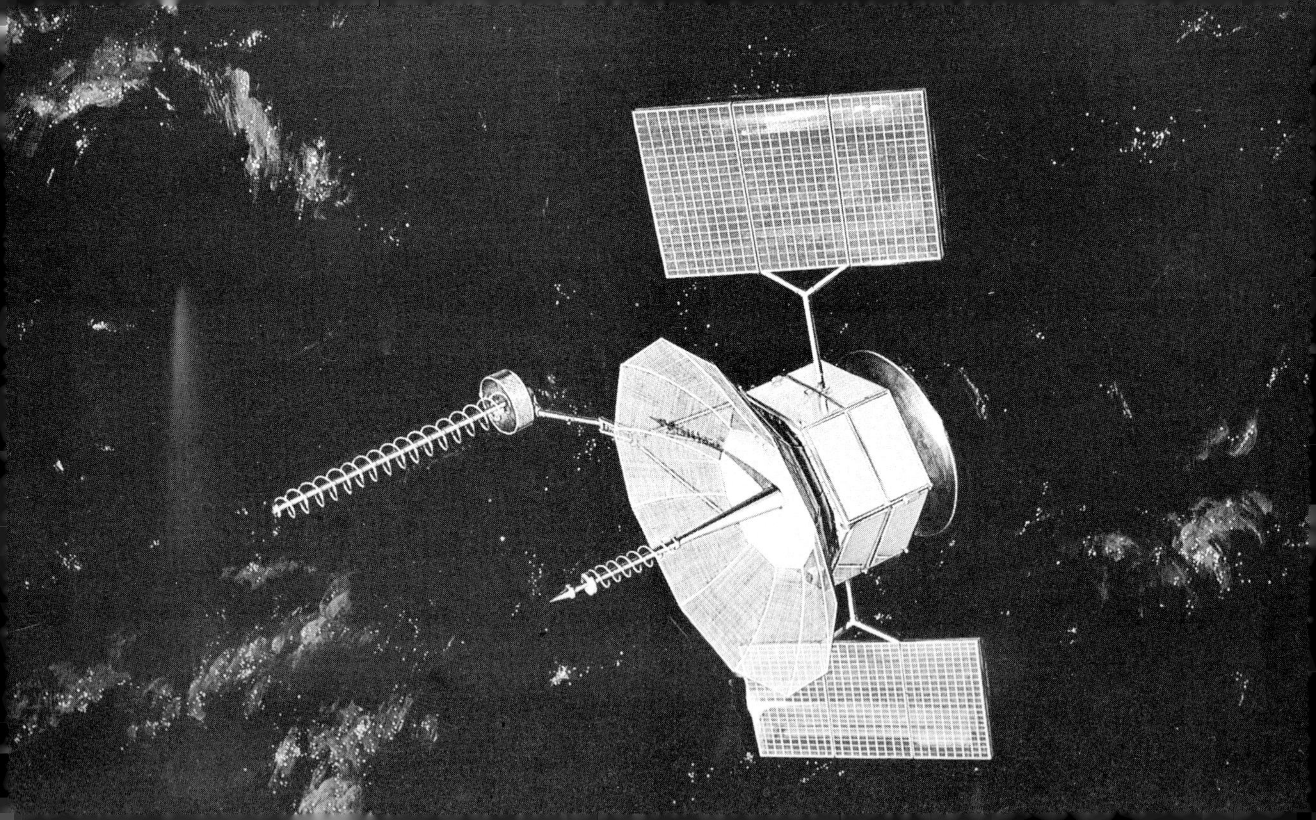

(USA — DOD)

Role: International communications
Data: Fltsatcom
Height: (body) 4 ft 2 in (1.27 m). Overall: 16 ft (4.9 m)
Diameter: 8 ft (2.4 m). Array span: 30 ft (9.1 m)
Orbital mass: 2 040 lb (927 kg)
Spacecraft: Three-axis stabilised geostationary satellite comprises hexagonal body carrying a pair of 7 ft x 13 ft (2.13 m x 3.96 m) deployable solar arrays, together generating more than 1.5 kW. Modular design splits body into lower spacecraft section containing control subsystems and solid-fuel apogee motor and upper section carrying payload. Design life is five years
Payload: Communications repeater using two transponders to provide at least 30 voice channels and 12 teletype channels for relay and communication with bases, submarines, ships and aircraft. One transponder uses UHF frequencies through 16 ft (2.44 m) deployable mesh antenna while the other provides SHF links through a small horn antenna projecting through side of UHF dish. TT&C uses S-band frequencies through central mast antenna
Programme: US Navy network using four identical spacecraft, possibly plus back-up unit, to be launched, beginning late 1975, into geostationary positions for global coverage. Prime contractor: TRW
Launch vehicle: Atlas Centaur

TACSAT

(USA — DOD)

Role: Communications technology
Data: Tacsat 1
Height: 25 ft (7.6 m) overall
Diameter: 9 ft (2.7 m)
Orbital mass: 1 600 lb (725 kg)
Spacecraft: Geostationary spacecraft has cylindrical main body surmounted by antenna assembly. Lower half of body spins at 54 rpm to provide stabilisation and contains apogee motor plus spacecraft controls. Upper half, containing communications subsystem and antennae, is despun. Dual spin technique permits longer body required to carry large area of solar array which covers both sections of body
Payload: Communications repeater is most powerful to date, transmitting at 6 kW effective radiated power. Antenna assembly includes two microwave horns for X-band plus set of five 8 ft (2.4 m)-long helical antennae providing high-gain narrow-beam signals to test use of two sub-bands in X-band — 225-400 MHz (UHF) and 7-8 gHz (SHF)
Programme: Tacsat 1 (originally Tacomsat) is single-spacecraft programme (to date) launched in February 1969 to test tactical communications by satellite — high-power space segment and small mobile Earth terminals. Tacsat is largest and most powerful comsat yet orbited and is second generation programme. Previous series of six Lincoln Experimental Satellites (LES) launched between February 1965 and September 1968 and varying in size from 69 lb to 360 lb (31 kg-163 kg) provided basic technology for both tactical and strategic communications. Tacsat (Pacific) and LES 6 (Atlantic) still operating in mid-1975. Prime contractor: Hughes
Launch vehicle: Titan IIIC

EOLE

Role: Meteorology
Data: Eole
Height: 23 in (0.58 m)
Diameter: 38 in (0.96 m)
Orbital mass: 186 lb (84 kg)
Spacecraft: Octagonal structure surmounted by spiral antenna is stabilised by deployable 33 ft (10.0 m) gravity-gradient boom. Eight small deployable solar panels attached round the base plus panels on each facet of main body, together generate 20 W of power. Design life is six months
Payload: Communications and data handling equipment. One 4 W transmitter interrogates balloons while another transmits data to Earth at 250 mW, both using PCM/PM techniques
Programme: Single spacecraft launched into 560-mile (900 km) circular orbit in August 1971 to relay data from 500 instrumented balloons released from Argentina over the following five months. Floating at 39,000 ft (12,000 m), balloons provided data on winds, temperatures and pressures in the southern hemisphere. Prime contractor: Aerospatiale
Launch vehicle: Scout B

RB211

(ESA)

METEOSAT

Role: Meteorology
Data: Meteosat
Height: 4 ft 9 in (1.45 m). Overall 10 ft 6 in (3.2 m)
Diameter: 6 ft 10 in (2.1 m)
Orbital mass: 776 lb (352 kg). Launch weight 1 537 lb (697 kg)
Spacecraft: Drum-shaped geostationary satellite spin-stabilised at 199 rpm. Body-mounted solar array generates 200 W. Design life is three years
Payload: Two-channel radiometer provides visible (day) imaging with 1½-mile (2.5 km) resolution and infra-red (night) imaging with 3-mile (5.0 km) resolution. Transmits a "full-disc" picture every 20 minutes. Also equipped to redisseminate pictures after ground processing and to relay data from remote data-collection platforms
Programme: Single satellite is planned for launch to geostationary orbit in 1977 to participate in the 1977 Global Atmospheric Research Programme along with US, Japanese and Russian metsats (see SMS). With technical assistance provided by Philco Ford, craft is quite similar to SMS. Prime contractor: Aerospatiale (for Cosmos consortium)
Launch vehicle: Delta 2914

GEOSTATIONARY METEOROLOGICAL SATELLITE (Japan — Nasda)

Role: Meteorology
Data: GMS
Height: Overall 10 ft 2 in (3.1 m)
Diameter: 7 ft 3 in (2.2 m)
Orbital mass: 772 lb (350 kg)
Spacecraft: Spin-stabilised satellite comprises shallow drum-shaped main body with apogee motor bell projecting below and main bulk of mission sensor projecting above. Twin helical aerials fitted on top provide constant picture transmission
Payload: Single two-channel radiometer for day and night viewing of cloud cover
Programme: Single spacecraft is scheduled for launch in 1977 into geostationary orbit to provide Japanese contribution to 1977 Global Atmospheric Research Programme (see Meteosat). Prime contractor: Nippon Electric Co
Launch vehicle: Delta 2914

(USA — Nasa)

NIMBUS

Role: Meteorological technology
Data: Nimbus 5
Height: 10 ft (3.05 m)
Diameter: 5 ft (1.52 m). Array span: 11 ft (3.35 m)
Orbital mass: 1 580 lb (717 kg). Launch weight: 1 695 lb (772 kg)
Spacecraft: Butterfly-shaped craft comprises circular lower section housing payload and hexagonal upper section containing three-axis stabilisation system and other control subsystems, the two being separated by an open truss. A pair of deployable solar panels generates more than 200 W
Payload: 490 lb (222 kg) package of experimental sensors for possible use in operational metsats. Includes improved versions of two instruments already Nimbus-tested — selective chopper radiometer and temperature humidity infra-red radiometer — plus microwave spectrometer, surface composition mapping radiometer, infra-red temperature profile radiometer and electrically scanning microwave radiometer
Programme: Testbed satellites for new meteorological sensors, systems and techniques plus active meteorological data collection. Early series of four spacecraft launched between August 1964 and April 1970. First of two Advanced Nimbus craft (Nimbus 5) launched December 1972. Nimbus 6 launched in June 1975. Another craft (Nimbus G) ordered for launch in 1978. Prime contractor: General Electric
Launch vehicle: Delta 900

TIROS OPERATIONAL SATELLITE

(USA —ESSA)

Role: Meteorology
Data: Essa 9
Height: 22 in (0.56 m)
Diameter: 42 in (1.07 m)
Orbital mass: 320 lb (145 kg)
Spacecraft: Drum-shaped 18-sided polygon covered with solar cells on all faces and spin-stabilised at 9.2 rpm. Single 18 in (46 cm) reception antenna projecting from top and four 22 in (56 cm) transmitting antennae extending from bottom
Payload: Two 1-in (2.5 cm) Advanced Vidicon Camera System (AVCS) cameras viewing through opposite sides of space craft provide constant daytime view of cloud formations
Programme: Nine similar spacecraft (Essa 2, 6 and 8 used different Automatic Picture Transmission cameras) launched into Sun-synchronous near-polar orbit between February 1966 and February 1969. Pictures received directly from APT satellites or via NASA centres from AVCS satellites by 400 stations in 45 countries including universities, radio stations and private individuals. Tiros Operational Satellites (TOS) financed by US Environmental Science Services Administration (ESSA) and redesignated after launch. This first operational system is based on technology from experimental Tiros (Television Infra-Red Observation Satellite) programme in which ten satellites were launched between 1960 and 1965. Prime contractor: RCA
Launch vehicle: T-A Improved Delta (TAID)

IMPROVED TIROS OPERATIONAL SATELLITE (TIROS-M)

Role: Meteorology
Data: NOAA 3
Height: 49 in (1.2 m)
Width: 40 in x 40 in (1.25 m x 1.25 m). Array span: 14 ft (4.3 m)
Orbital mass: 750 lb (340 kg)
Spacecraft: Box-shaped craft has three deployable solar arrays projecting from top and generating 250 W. Three-axis stabilised by large spinning flywheel controlling pitch. Real-time data transmission on S-band. Design life is one year
Payload: Four instruments include scanning radiometer with visible channel for daylight scans of surface to 2.3 miles (3.7 km) resolution and infra-red channel for day and night pictures of cloud cover. Another radiometer scans smaller areas at ½-mile resolution. I-r radiometer measures at six atmospheric levels and at surface while a solar proton monitor provides warning of solar storms. Payload similar to that of NOAA-2 — first Itos without cameras
Programme: Second generation metsat, named Tiros M, has greatly increased capability over predecessor. First in series, Itos 1, provided by Nasa as R & D model and launched into high, near-polar orbit in January 1970. National Oceanic and Atmospheric Administration formed in 1970 under US Commerce Department, took over most of ESSA's responsibilities. Itos 2 redesignated NOAA 1. NOAA 2, 3 and 4, plus two launch failures, completed programme by November 1974. Satellites financed and now fully operated by NOAA. Itos 1 and NOAA 1 and 4 have different payload including pairs of AVCS and APT cameras (see TOS). Three more craft, Itos I, J and K now on order. Third generation system, Tiros N, also being studied. Prime contractor: RCA
Launch vehicle: Two-stage Delta

SYNCHRONOUS METEOROLOGICAL SATELLITE (GOES) (USA — NOAA)

Role: Meteorology
Data: SMS 1
Height: 8 ft 7 in (2.62 m). Overall: 11 ft 4 in (3.45 m)
Diameter: 6 ft 3 in (1.9 m)
Orbital mass: 535 lb (243 kg). Launch weight: 1 379 lb (627 kg)
Spacecraft: Geostationary drum-shaped satellite spin-stabilised at 100 rpm. Body-mounted solar array generates 200 W. 4,900 lb (21.8 kN)-thrust apogee motor. Design life is three-five years
Payload: Day and night data collection provided by Visible Infra-red Spin-Scan Radiometer (looking through side aperture) giving I-R night resolution of 5 miles (9 km) and visible day resolution of ½ mile (0.9 km). A "full disc" picture of the Western hemisphere is transmitted every 30 min. Also carried is Space Environment Monitoring (SEM) package which provides data on solar flare activity, incoming X-radiation and intensity of Earth's magnetic field
Programme: World's first geostationary metsat is Nasa-financed craft launched in May 1974 and then turned over to NOAA after international meteorological exercise and relocated at 75° W (see Itos). SMS 2 launched in January 1975 to cover E hemisphere and both will participate in the 1977 Global Atmospheric Research Programme in conjunction with Europe's Meteosat, a Japanese metsat and Soviet geostationary Meteor. A third SMS was to be launched in October 1975 for NOAA and redesignated GOES-A (for Geostationary Operational Environmental Satellite) for which SMS 1 and 2 were originally considered as R & D craft. Two more SMSs ordered for launch as GOES B and C in 1976 and 1977. GOES will also relay data from 10,000 land and sea data-collection platforms around USA. Prime contractor: Philco Ford
Launch vehicle: Delta 2914

DEFENCE METEOROLOGICAL SATELLITE PROGRAMME

Role: Meteorology
Data: Block 5D DMSP
Height: 17 ft (5.2 m)
Diameter: 6 ft (1.8 m)
Orbital mass: 1 043 lb (473 kg). Launch weight: 5 900 lb (2 676 kg)
Spacecraft: Structurally divided into three elements arranged in tandem comprising precision-pointing sensor platform on top, electronics section in middle and third stage rocket casing enclosed by attitude control system at base. Single, steerable solar array generates 900 W. Three-axis stabilisation provides 0.1° pointing accuracy. Unique system of two processors automatically controls both launcher and spacecraft. Four pitch/yaw jets provide 155 lb (689 N) thrust for first stage control and separation, 65 lb (289 N) thrust for second stage control and separation. Also for terminal manoeuvres but switching to nitrogen cold-gas system
Payload: 300 lb (136 kg) package of seven sensors provides temperature and water vapour measurements and cloud-top imagery at 1,700 ft (500 m) resolution. Localised data stored for transmission to USAF centres and provided in real time to field centres
Programme: Second-generation DOD operational metsat now under development will replace current RCA network of 500 lb (227 kg) satellites (details classified) possibly consisting of six launches between October 1971 and end of 1974. Preceded by experimental systems comprising small satellites launched at a rate of two to three per year. Prime contractor: RCA

METEOR

Role: Meteorology
Data: Meteor
Height: 16 ft (4.88 m) approx
Diameter: 5 ft (1.5 m)
Spacecraft: Three-axis stabilised craft comprising main cylindrical body with smaller sensor compartment at lower end and two Sun-orientating solar arrays
Payload: Infra-red radiometer covering a 1,550-mile (2,500 km) wide path and imaging television covering a 620-mile (1,000 km) wide path giving 24 h data and claimed to have a resolution three times better than that of US Tiros metsats. Data storage and real-time transmission facilities carried
Programme: Ongoing series of improved or replacement satellites maintaining basic three-satellite phased network in polar orbit. Since first launch in March 1969, 20 Meteors had been launched by end of 1974. Preceded by R & D programme of 14 flights under Cosmos label most being interpreted as such (including Cosmos 23, 44, 45, 65 and 92) and the rest declared as such (Cosmos 122, 144, 156, 184, 206 and 226). Meteors are stated to be part of a three-level approach including Salyut observations and future geostationary metsats. A geostationary metsat is anticipated for the Global Atmospheric Research Programme in 1977 (see SMS)
Launch vehicle: A-1

B-1 COSMOS (USSR)

Data: B-1 Cosmos
Configuration: Two-stage light launch vehicle for civil and military payloads. Third stage sometimes used. First stage is modified Sandal IRBM
Height: Overall 105 ft 4 in (32.1 m). First stage 66 ft 7 in (20.3 m). Second stage 27 ft 11 in (8.5 m). Shroud 10 ft 10 in (3.3 m)
Diameter: 5 ft 4 in (1.6 m) overall except for base skirting and fins
Launch weight: First stage about 59 500 lb (27 000 kg)
First stage propulsion: Single RD-214 engine with four chambers. Propellants nitric acid/kerosene. Thrust (s.1.) 160 965 lb (715 kN)
Second stage propulsion: RD-119 single chamber plus pitch, roll and yaw vernier thrusters. Propellants lox/UDMH. Thrust (vac) 24 250 lb (107.8 kN)
Performance: About 900 lb (410 kg) into low orbit at 48°
Typical missions: Cosmos (military and civil), Intercosmos
Operational record: Entered service March 16, 1962. More than 120 apparently successful launches by end of 1974. Launcher still in service

Data: C-1
Configuration: Light military launcher of two or three stages. First stage is modified SS-5 Skean IRBM.
Height: About 105 ft 6 in (32.2 m). First stage 62 ft 7 in (19.1 m)
Diameter: Overall 8 ft 2 in (2.5 m)
Guidance and control: Inertially guided using first stage fins and possibly other systems for first stage
Performance Up to 2 000 lb (900 kg) in low orbit
Typical missions: Multiple payload navigation and ferret. Some Cosmos, Intercosmos
Operational record: Introduced in August 1964. Achieved at least 97 successful flights by end of 1974. Still in service

A-1 VOSTOK (USSR)

Data: A-1 Vostok
Configuration: Two and-a-half-stage medium launcher comprises two-stage core plus four strap-on boosters. Lower core and strap-ons modified from SS-6 Sapwood ICBM or basic A vehicle
Height: 124 ft 7 in (38.0 m). Boosters 62 ft 4 in (19.0 m). First stage 90 ft 2 in (27.5 m). Second stage 11 ft 2 in (3.4 m). Shroud 21 ft 7 in (6.6 m)
Diameter: Overall (base) 33 ft 9 in (10.3 m). Booster 10 ft (3.0 m). First stage 9 ft 7 in (2.95 m). Upper stage 8 ft 6 in (2.6 m). Shroud 8 ft 9 in (2.8 m)
Booster propulsion: Four boosters each with single RD 107 engine comprising four thrust chambers and two verniers. Propellants lox/kerosene. Thrust 224 910 lb (1 000.4 kN) per booster. Isp 314 sec
First stage propulsion: Single RD 108 engine comprising four thrust chambers and four verniers. Propellants lox/kerosene. Thrust 224 910 lb (1 000.4 kN)
Second stage propulsion: Single-chamber engine, designation and propellant unannounced. Thrust (vac) 198 450 lb (822.7 kN) implied in announced 600 tonnes total thrust for vehicle
Guidance and control: Radio guidance using verniers for first stage, thrusters for second stage
Performance: 10 400 lb (4 720 kg) into low orbit, 1 000 lb (450 kg) to Earth escape
Typical missions: Vostok, Cosmos, Luna 1 — 3
Operational record: Introduced without upper stage as Sputnik 1 launcher in October 1957. Used successfully with various upper stages for at least 482 orbital, interplanetary and manned launches by end of 1974 (see A-2). Still in service

A-2 SOYUZ

Data: A-2 Soyuz (SL-4)
Configuration: Two-and-a-half stage medium launcher has two-stage core plus four strap-on boosters. Lower core stage and strap-ons developed from SS-6 Sapwood ICBM. Extra stage added for Earth escape version (A-2-e)
Height: Overall 167 ft 4 in (51 m). Boosters 62 ft 4 in (19.0 m). First stage 90 ft 2 in (27.5 m). Second stage 30 ft 9 in (30.8 m). Shroud/escape rocket 44 ft 7 in (13.6 m)
Diameter: Overall (base) 33 ft 9 in (10.3 m). Boosters 10 ft (3 m) each. First stage 9 ft 7 in (2.95 m). Second stage 8 ft 6 in (2.6 m). Shroud 8 ft 9 in (2.8 m)
Launch weight: 720 000 lb (325 592 kg)
Booster propulsion: Four boosters each with single RD 107 engine comprising four thrust chambers and two verniers. Propellants lox/kerosene. Thrust 224 868 lb (1,000.2 kN) each booster
First stage propulsion: Single RD 108 engine comprising four thrust chambers and four verniers. Propellants lox/kerosene. Thrust 211 640 lb (941.4 kN)
Second stage propulsion: Four-chamber engine, designation and propellants unannounced. Propellants lox/kerosene. Thrust (vac) 66 138 lb (294.2 kN)
Guidance and control: radio guidance using verniers for first stage, thrusters for second stage
Performance: 16 550 lb (7 500 kg) into low orbit, 3 640 lb (1 650 kg) to Earth escape (A-2-e)
Typical missions: Voskhod, Soyuz, Cosmos, Luna, Venera, Mars
Operational record: Introduced with escape stage in October 1960. First basic A-2 version launched November 1963 and is most widely used A variant. Still in service

Data: D-1 Proton (SL-13), F-1
D-1 description: Purpose-built heavy launcher for civil use has three stages plus fourth for Earth escape missions (D-1-e). First stage thrust estimated at about 3.4 million lb (15 123 kN). Introduced July 1965, Still in service
Performance: At least 37 585 lb (17 000 kg) into low orbit, at least 12 000 lb (5 440 kg) to Earth escape (D-1-e). Typical missions include Proton, Salyut, Zond 4-8, Luna 15/16, Mars. At least 40 successful flights to orbit by end of 1974
F-1 description: Intermediate military launch vehicle of four stages. First and second stages are SS-9 Scarp ICBM. First stage height 67 ft 11 in (20.7 m), diameter 9 ft 10 in (3 m). Second stage height 38 ft 4 in (11.7 m), diameter 3 ft 8 in (1.1 m). Overall height estimated about 155 ft (47.2 m). First stage engine system has six nozzles plus four verniers. Introduced September 1966. Still in service
Performance: At least 10 000 lb (4 550 kg) into low orbit. Missions include Cosmos (FOBS, Inspector). At least 40 successful flights to orbit by end of 1974

BLACK ARROW

(UK)

Data: Black Arrow
Configuration: Three-stage vehicle with third stage enclosed within shroud
Height: Overall 42 ft 9 in (13.0 m). First stage 19 ft 7 in (6.0 m). Second stage 11 ft 7 in (3.6 m). Third stage 3 ft 3 in (1.0 m). Shroud 11 ft 7 in (3.6 m)
Diameter: First stage 6 ft 7 in (2.0 m). Second stage 4 ft 6 in (1.4 m). Third stage 2 ft 4 in (0.7 m). Shroud 4 ft 6 in (1.4 m)
Launch weight: Overall 40 000 lb (18,144 kg). First stage 31 338 lb (14 215 kg). Second stage 7 826 lb (3 550 kg). Third stage 778 lb (353 kg)
Empty weight: First stage 2 238 lb (1 015 kg). Second stage 816 lb (370 kg). Third stage 77 lb (35 kg)
First stage propulsion: Single Rolls-Royce Gamma Type 8 engine with eight nozzles. Propellants hydrogen peroxide/kerosene. Thrust (s.1.) 52 075 lb (227 kN). Burn time 125 sec
Second stage propulsion: Single Rolls-Royce Gamma type 2 engine with two nozzles. Propellants hydrogen peroxide/kerosene. Thrust (vac) 15 750 lb (70 kN). Burn time 121 sec
Third stage propulsion: Bristol Aerojet Waxwing motor. Solid propellant. Thrust 4 927 lb (21.9 kN). Burn time 55 sec
Guidance and control: Inertially guided using gimballed nozzles on first two stages. Third stage spin-stabilised
Performance: 220 lb (100 kg) into 300 mile (480 km) polar orbit
Typical mission: X-3
Operational record: Programme cancelled after single operational launch in November 1971
Prime contractor: Westland Aircraft

DIAMANT B-P4

Data: Diamant B-P4
Configuration: Three-stage light launcher, shortened and uprated by new second stage originally developed for M-1 MSBS
Height: 71 ft (21.6 m). First stage 49 ft (14.9 m). Second stage 7 ft 6 in (2.3 m). Third stage 5 ft 6 in (1.7 m). Shroud 13 ft 5 in (4.1 m)
Diameter: First stage 4 ft 7 in (1.4 m). Second stage 4 ft 11 in (1.5 m). Third stage 2 ft 7 in (0.8 m)
Launch weight: 60 610 lb (27 492 kg). First stage 45 300 lb (20 550 kg). Second stage 10 538 lb (4,780 kg). Third stage 1 720 lb (780 kg)
Empy weight: First stage 4 850 lb (2 200 kg). Second stage 1 720 (780 kg), third stage 209 lb (95 kg)
First stage propulsion: (L-17) SEP Valois engine. Propellants nitrogen tetroxide/UDMH. Thrust (s.1) 78 300 lb (348 kN). Burn time 119 sec
Second stage propulsion: (P-4) SEP Rita 1 engine. Solid propellants. Thrust 40 500 lb (180 kN). Burn time 55 sec.
Third stage propulsion: (PO6) SEP engine. Solid propellant. Thrust 11 250 lb (50 kN). Burn time 46 sec.
Performance: 440 lb (200 kg) into 186 mile (300 km) orbit
Operational record: Initial Diamant A version successfully orbited payloads on all four missions from December 1965 to 1967. Diamant B introduced in March 1970 and flown five times with two failures preceding B-P4 replacement. Diamant service ended after three B-P4 launches during 1975
Prime contractor: Aerospatiale

EUROPA II

Data: Europa II
Configuration: Four-stage light civil launcher. First stage is modified Blue Streak IRBM. Fourth stage enclosed by shroud
Height: 103 ft (31.6 m). First stage 64 ft 4 in (19.6 m). Second stage 18 ft (5.5 m). Third stage 12 ft 6 in (3.8 m). Fourth stage 6 ft 6 in (2.0 m). Shroud 13 ft (4.0 m)
Diameter: First stage: 10 ft (3.1 m). Second/third stage 6 ft 6 in (2.0 m). Fourth stage 2 ft 7 in (0.8 m)
Launch weight: 246 000 lb (111 586 kg). First stage 208 000 lb (94 350 kg). Second stage 26 455 lb (12 000 kg). Third stage 8 820 lb (4 000 kg). Fourth stage 1 764 lb (800 kg)
Empty weight: First stage 14 330 lb (6 500 kg). Second stage 5 952 lb (2 700 kg). Third stage 1 951 (885 kg). Fourth stage 231 lb (105 kg).
First stage propulsion: (Blue Streak) Rolls Royce RZ-2 two-chambered engine with gimballed nozzles. Propellants lox/kerosene. Thrust 300 000 lb (1 334.4 kN). Burn time 160 sec
Second stage propulsion: (Coralie) LRBA engine with four chambers. Propellants nitrogen tetroxide/UDMH. Thrust 59 520 lb (265 kN). Burn time 103 sec.
Third stage propulsion: (Astris) Erno engine with single nozzle and two verniers, Propellants nitrogen tetroxide/aerozine 50. Thrust 5 070 lb (22.5 kN) plus 90 lb (0.4 kN) from each vernier. Burn time 36 sec.
Fourth stage propulsion: SEP P 6 Diamant engine. Solid propellant. Thrust 9 260 lb (41.2 kN). Burn time 45 sec
Guidance and control: Inertial guidance using gimballed nozzles on first two stages, verniers on third stage
Performance: 800 lb (363 kg) into circular orbit, 440 lb (200 kg) into geostationary transfer orbit
Operational record: Programme cancelled after single development launch in November 1971 which was failure. Preceded by ten Europa 1 development launches between 1964 and 1970
Prime contractors: HSD (first stage), Aerospatiale (second stage), ASAT (third stage).

L-IIIS ARIANE

(ESA)

Data: L-IIIS Ariane
Configuration: Three-stage medium launcher vehicle
Height: 155 ft 6 in (47.4 m). First stage 60 ft 4 in (18.4 m). Second stage 37 ft 8 in (11.5 m). Third stage 28 ft 2in (8.6 m). Shroud 28 ft 5 in (8.6 m)
Diameter: First stage 12 ft 6 in (3.8 m). Second/third stage 8 ft 6 in (2.6 m). Shroud 10 ft 6 in (3.2 m)
Launch weight: 443 125 (201 000 kg). First stage 339 508 lb (154 000 kg). Second stage 81 570 lb (37 000 kg). Third stage 19 840 lb (9 000 kg)
Empty weight: First stage 29 255 lb (13 270 kg). Second stage 7 143 (3 240 kg). Third stage 2 491 lb (1 130 kg). Shroud 1 760 lb (800 kg)
First stage propulsion: (L 140) Four SEP Viking II engines linked to single gimballed mounting. Propellants nitrogen tetroxide/UDMH Thrust, (s.1.) 133 144 lb (592.2 kN) each. Burn time 140 sec
Second stage propulsion: (L 33) Single gimbal-mounted Viking II engine. Propellants nitrogen tetroxide/UDMH. Thrust (vac) 157 275 lb (699 kN). Burn time 131 sec
Third stage propulsion: (H 8) Single SEP/MBB HM-7 engine. Propellants lox/liquid hydrogen. Thrust 13 500 lb (60 kN). Burn time 563 sec
Performance: 5 500 lb (2 500 kg) in 620 mile (1 000 km) circular orbit, 3 300 lb (1 500 kg) in geostationary transfer orbit, 2 100 lb (950 kg) to Earth escape
Operational record: Launcher is under collaborative development in Europe for European Space Agency. Replaces cancelled Europa III programme. Ariane due to enter service in 1980
Prime contractor: Aerospatiale

Data: Mu-3C
Configuration: Three-stage all solid-propellant light launch vehicle. First stage has clustered strap-on boosters. Third stage enclosed in shroud
Height: Overall 66 ft 5 in (20.2 m). Boosters 19 ft (5.8 m). First stage 39 ft 1 in (11.9 m). Second stage 16 ft (4.9 m). Third stage 5 ft 1 in (1.5 m). Shroud 11 ft 9 in (3.6 m)
Diameter: Overall (core section) 4 ft 7 in (1.41 m). Fin span 14 ft 9 in (4.5 m). Boosters 1 ft (0.31 m). Third stage 3 ft 9 in (1.14 m)
Launch weight: Overall 93 050 lb (42 207 kg). First stage 55 423 lb (25 140 kg). Boosters 1 054 lb (478 kg) each. Second stage 19 038 lb (8 626 kg). Third stage 2 736 lb (1 241 kg)
First stage propulsion: M-10 engine. Thrust (s.1.) 194 670 lb (865.9 kN). Boosters 30 750 lb (136.8 kN) each. Total lift-off thrust 440 670 lb (1 960.1 kN)
Second stage propulsion: M-22 engine. Thrust (vac) 80 910 lb (359.9 kN)
Third stage propulsion: M-3A engine. Thrust 16 535 lb (73.5 kN)
Guidance and control: Inertial guidance for first stage using freon-injection thrust-vectoring plus vernier thrusters. Third stage radio-guided
Performance: 165 lb (75 kg) into 217 x 373 mile (350 x 600 km) orbit
Typical missions: Tansei II, Taiyo
Operational record: Introduced as four-stage vehicle with four launches between September 1970 and August 1972, including failure of first flight. Mu-3C variant used for three missions in 1974/1975. Two further missions scheduled in 1976 and 1977 of four-stage versions with improved guidance
Prime contractor: Nissan

N VEHICLE

(Japan — Nasda)

Data: N

Configuration: Three-stage medium launcher. First stage is licence-built version of modified Thor IRBM. Third stage enclosed in shroud

Height: Overall 106 ft 11 in (33.0 m). First stage 70 ft 3 in (21.4 m). Second stage 18 ft (5.5 m). Shroud 18 ft 8 in (5.7 m)

Diameter: First stage 8 ft (2.4 m). Second stage 4 ft 7 in (1.4 m). Third stage 3 ft 11 in (1.2 m). Shroud 5 ft 5 in (1.6 m)

Launch weight: Overall 198 000 lb (90 000 kg). First stage 154 476 lb (70 070 kg). Boosters (each) 9 832 lb (4 460 kg). Second stage 12 610 lb (5 720 kg). Third stage 1 455 lb (660 kg).

Empty weight: First stage 9 634 lb (4 370 kg). Boosters (each) 1 657 lb (760 kg). Second stage 2 249 lb (1 020 kg). Third stage 159 lb (72 kg)

First stage propulsion: Mitsubishi-built Rocketdyne MB-3-3 engine. Propellants lox/kerosene. Thrust (s.1.) 170 000 lb (756.1 kN). Burn time 218 sec. Three Thiokol Castor II boosters. Solid propellant. Thrust 52 000 lb (231.3 kN) each. Burn time 40 sec

Second stage propulsion: Mitsubishi LE-3 engine. Propellants nitrogen tetroxide/aerozine 50. Thrust (vac) 12 000 lb (53 kN). Burn time 246 sec

Third stage propulsion: Thiokol TEM364-14 motor. Solid propellant. Thrust 8 708 lb (387.3 kN). Burn time 40 sec

Guidance and control: Inertial guidance using first stage gimballed verniers, second stage gimballed verniers and hydrazine thrusters

Performance: 2 200 lb (1 000 kg) into 125 mile (200 km) orbit, 287 lb (130 kg) into geostationary transfer orbit

Typical missions: ETS, ISS

Operational record: New Japanese launcher made its first flight in September 1975 launching ETS 1. Three more missions scheduled by 1977

Prime contractor: Mitsubishi

SCOUT

Data: Scout D
Configuration: Four-stage, all solid-propellant light launcher. Fourth stage enclosed within high-volume shroud. Can take optional Alcyone 1A fifth stage (Scout E)
Height: Overall 73 ft (22.2 m).First stage 30 ft (9.1 m). Second stage 20 ft (6.1 m). Third stage 10 ft (3.0 m). Fourth stage 6 ft (1.8 m). Shroud 13 ft (3.0 m)
Diameter: First stage 3 ft 9 in (1.15 m). Second/third stage 2 ft 6 in (0.75 m). Fourth stage 1 ft 8 in (0.5 m)
Launch weight: Overall 47 000 lb (21 546 kg). First stage 30 000 lb (13 605 kg). Second stage 9 743 lb (4 410 kg). Fourth stage 660 lb 299 kg)
First stage propulsion: UTC Algol IIIA. Thrust 140 000 lb (622.3 kN). Burn time 82 sec
Second stage propulsion: Thiokol Castor IIA. Thrust 60 000 lb (266.7 kN). Burn time 40 sec
Third stage propulsion: Hercules X-259 Antares IIB. Thrust 21 000 lb (93.3 kN). Burn time 40 sec
Fourth stage propulsion: UTC FW-4S Altair III. Thrust 5 620 lb (25 kN). Burn time 31 sec
Guidance and control: Inertially guided. First stage has thrust-vector vanes and tail fins. Second and third stages have hydrogen peroxide jets. Fourth stage spin-stabilised
Performance: 390 lb (177 kg) into 300 mile (480 km) orbit
Typical missions: UK-4, X-4, small Explorers, Transit (NNS)
Operational record: Entered service in February 1961. Launches total 88 missions with 83 successes by end of 1974. Still in service
Prime contractor: LTV Aerospace

UNITED STATES

THOR DELTA

Data: Delta 1904
Configuration: Three-stage medium launch vehicle with three to nine strap-on boosters. First stage is extended long-tank Thor
Height: Overall 116 ft (35.3 m). First stage 70 ft (21.3 m). Second stage 21 ft (6.4 m). Third stage 4 ft 6 in (1.4 m). Shroud 26 ft (7.9 m)
Diameter: First stage 8 ft (2.4 m). Second stage 5 ft (1.5 m). Third stage 3 ft (0.9 m). Shroud 5 ft 5 in (1.65 m)
Launch weight: Overall 290 000 lb (131 500 kg). First stage 186 000 lb (84 370 kg). Second stage 13 690 lb (6 210 kg). Third stage 1 592 lb (722 kg)
First stage propulsion: Rocketdyne MB-3-3 engine. Propellants lox/kerosene. Thrust (s.1.) 175 000 lb (778.4 kN). Burn time 228 sec. Nine strap-on Thiokol Castor II solid-fuel boosters, six igniting at launch, three after 39 sec. Thrust 52 000 lb (231 kN) each. Burning time 38 sec.
Second stage propulsion: Aerojet AJ-10-118F engine. Propellants nitrogen tetroxide/aerozine 50. Thrust (vac) 9 500 lb (42.3 kN). Burn time 335 sec
Third stage propulsion: Thiokol TE-364-4 engine. Solid propellant. Thrust 13 900 lb (61.8 kN). Burn time 44 sec
Guidance and control: Delta Inertial Guidance System uses first stage gimballed nozzle, second stage thrusters
Performance: 1 500 lb into geostationary transfer orbit, 4 050 lb (1 837 kg) into 230 miles (370 km) orbit (two-stage)
Typical missions: AE-C. IMP, Geos
Operational record: Delta 1 000 series also has Straight Eight version and entered service in September 1972. Three flights, all successful, by end of 1974. Still in service
Prime contractor: McDonnell Douglas

TeleSat
Canada

DELTA "STRAIGHT EIGHT"

Data: Delta 2914
Configuration: Latest Delta variant has upper stages contained within constant diameter structure to increase payload volume. Variants include two-stage model, variation in strap-on solid rocket motors (SRMs) from three to nine. (Four-figure designator shows configuration). First stage is developed from Thor IRBM
Height: 116 ft (35.4 m) including shroud. Strap-on solids 23 ft 7 in (7 m). First stage 70 ft (21.3 m) with new 10 ft (3 m) tank extension. Second stage 21 ft (6.4 m). Third stage 4 ft 6 in (1.4 m). Shroud 26 ft (7.9 m)
Diameter: 8 ft (2.4 m) overall without strap-on solids. Solids 31 in (0.8 m) each. Second stage: 5 ft (1.5 m). Third stage 3 ft (0.9 m)
Launch weight: Overall 293 000 lb (133 000 kg). Strap-on solids 9 850 lb (4 468 kg) each. First stage 186 000 lb (84 370 kg). Second stage 13 600 lb (6 180 kg). Third stage 2 560 lb (1 160 kg)
First stage propulsion: Single Rocketdyne RS-27 (modified H-1) engine gimbal-mounted for pitch and yaw control and with two vernier jets for roll. Propellants lox/kerosene. Thrust 205 000 lb (911.2 kN). Burn time 228 sec. Nine strap-on Thiokol Castor II solid-fuel boosters, six igniting at launch, three after 39 sec. Thrust 52 000 lb (231 kN) each. Burn time 38 sec
Second stage propulsion: Single TRW TR-201 engine modified from Lunar Module descent engine. Propellants nitrogen tetroxide-aerozine 50. Thrust 9 500 lb (42.2 kN). Burn time 335 sec. Gas-jet attitude control
Third stage propulsion: Single Thiokol TE-364-4 with spin table. Solid propellant. Thrust 13 900 lb (61.8 kN). Burn time 44 sec
Guidance and control: Inertial guidance for gimballed nozzle plus two vernier jets on first stage and gimballed nozzle on second stage
Performance: 4,500 lb (1 920 kg) into low orbit, 1,500 lb (680 kg) into geostationary transfer orbit, 850 lb (386 lb) to Earth escape
Typical missions: SMS, Anik, Westar. (Two-stage) Skynet, Itos
Operational record: Eleventh major variant since May 1960. Entered service in November 1972 with Delta 92. Ten missions by end of 1974 with one failure. Constantly uprated Delta series, totalled 106 launches by end of 1974 (ten failures)
Prime contractor: McDonnell Douglas

ATLAS

(USA — DOD)

Data: Atlas F
Configuration: One-and-a-half-stage launcher is refurbished Atlas ICBM
Height: 71 ft 3 in (21.7 m). Up to 87 ft (26.5 m) with payload shroud
Diameter: 10 ft (3 m)
Launch weight: 269 000 lb (122 000 kg)
First stage propulsion: Single Rocketdyne MA-3 one-and-a-half-stage system comprising two boosters, jettisoned in flight, flanking one sustainer engine, plus two vernier engines. All engines fed by single propellant supply of lox/kerosene. Thrust total 389 000 (1 730 kN), sustainer 57 000 lb (253.5 kN), boosters 165 000 lb (734 kN) each, verniers 1 000 lb (4.45 kN)
Guidance and control: Inertially guided using verniers
Performance: 2 000 lb (907 kg) into 115-mile (185 km) orbit, 6 000 lb (2 720 kg) on 5 750 mile (9 265 km) ballistic flight
Typical missions: Re-entry vehicles, OV-1, Mercury capsules (Atlas D)
Operational record: Atlas B, D, E and F used as launcher since November 1959, usually with Agena A and B upper stages (also Able and Burner II). Some still remaining (E and F only). Including purpose-built SLV-3 series, total of Atlas space launch flights was 159 by end of 1974
Prime contractor: General Dynamics Convair

ATLAS-AGENA

Data: Atlas-Agena D
Configuration: Two-and-a-half-stage general purpose launcher comprising specially manufactured Atlas SLV-3A plus Agena-D upper stage
Height: Overall up to 120 ft (36.6 m). First stage 71 ft (21.6 m). Second stage 22 ft (6.7 m). Shroud up to 27 ft (8.2 m)
Diameter: First stage 10 ft (3.0 m). Second stage 5 ft (1.5 m)
Launch weight: Overall 325 637 (147 710 kg) without payload. First stage 310 600 (140 890 kg). Second stage 15 037 lb (6 821 kg)
First stage propulsion: One-and-a-half-stage Rocketdyne MA-5 system comprising two boosters flanking one sustainer engine and jettisoned in flight, plus two vernier engines. All fed by single propellant supply of lox/kerosene. Thrust total 431 300 lb (1 918 kN), sustainer 60 000 lb (266.9 kN), boosters 185 000 lb (822.9 kN) each, verniers 650 lb (2.9 kN) each. Burn time 139 sec (boosters) 247 sec (sustainer)
Second stage propulsion: see Agena D
Guidance and control: First stage — radio control using verniers. Second stage — inertial guidance using gas jet system
Performance: 8 800 lb (3 992 kg) into 115 mile (185 km) orbit, 2 920 lb (1 325 kg) into geostationary transfer orbit
Typical missions: Agena setellites, OGO, Lunar Orbiter, Mariner
Operational record: Atlas-Agena entered service in February 1960. Most heavily used Atlas application, predominantly by DOD. SLV-3A-Agena D introduced in March 1968 with five launches by end of 1974. Still in service
Prime contractor: General Dynamics Convair. Lockheed (Agena stage)

UNITED STATES
26

ATLAS — CENTAUR

Data: Atlas-Centaur
Configuration: Two-and-a-half-stage medium launcher comprises Atlas SLV-3D as first stage plus Centaur D-1A upper stage
Height: 131 ft (39.9 m). First stage 70 ft (21.3 m). Second stage 30 ft (9.14 m). Shroud 31 ft (9.4 m)
Diameter: Overall 10 ft (3 m)
Launch weight: Overall up to 325 732 lb (147 750 kg). First stage 283 600 lb (128 639 kg). Second stage 39 040 lb (17 708 kg)
First stage propulsion: One-and-a-half-stage MA-5 system comprising one sustainer engine flanked by two boosters jettisoned in flight, and two verniers. Propellants lox/kerosene (single supply for complete engine system). Thrust total 431 040 lb (1 917 kN), sustainer 60 000 lb (266.8 kN), boosters 185 000 (822.9 kN) each, verniers 650 lb (2.9 kN) each. Burn time 139 sec (boosters), 247 sec (sustainer). Velocity at sustainer cut-out 8 096 mph (13,029 km/h)
Second stage propulsion: see Centaur D-1T
Guidance and control: First stage — autopilot to booster cut-out using verniers. Sustainer phase and second stage — inertial guidance using 14 hydrogen peroxide thrusters on Centaur
Performance: 11 200 lb (5 080 kg) into 115 mile (185 km) orbit, 4 000 lb (1 814 kg) into geostationary transfer orbit, 1 300 lb (590 kg) to Earth escape
Typical missions: OSO, Intelsat IV, Surveyor, Mariner, Pioneer 10
Operational record: Entered service in May 1966. Total of 22 successes in 26 launches by end of 1974. Still in service
Prime contractor: General Dynamics Convair

UNITED STATES

TITAN-CENTAUR

Data: Titan IIIE-Centaur
Configuration: Four-stage heavy launcher comprises Centaur D-1T mated to Titan IIIE (two core stages plus "zero stage" boosters). Optional fifth stage for high velocity missions (Burner II). Centaur is enclosed with payload inside standard shroud.
Height: Overall 160 ft (48.8 m). Zero stage 85 ft (25.9 m). First stage 72 ft 11 in (22.2 m). Second stage 23 ft 4 in (7.1 m). Interstage 9 ft 6 in (2.9 m). Third stage 30 ft (9.1 m). Shroud 58 ft (17.6 m)
Diameter: Zero stage 10 ft (3.0 m) each. Core stage and Centaur 10 ft (3.0 m). Shroud 14 ft (4.2 m)
Launch weight: Overall 1 414 000 lb (641 400 kg). Titan 1 370 000 lb (621 400 kg). Centaur 35 843 (16 258 kg)
Zero stage propulsion: Two UTC 120 in (3.0 m) boosters. Solid propellant. Thrust 1.2 million lb (5 338 kN) each. Burn time 126 sec
First stage propulsion: Single Aerojet YLR87-AJ-11 engine with two chambers (gimballed). Propellants: nitrogen tetroxide/aerozine 50. Thrust (vac) 520 000 (2 313 kN). Burn time 146 sec
Second stage propulsion: Single-chamber Aerojet YLR91-A-11 engine. Propellants nitrogen tetroxide/aerozine 50. Thrust (vac) 101 000 lb (449 kN)
Third stage propulsion: see Centaur D-1T. Propellants: lox/liquid hydrogen. Thrust 15 000 lb (66.7 kN) each. 12 hydrogen peroxide thrusters
Guidance and control: Centaur inertial guidance using zero stage thrust-vectoring, first stage gimballed nozzles and Centaur thrusters
Performance: 38 000 lb (17 160 kg) into 115 mile (185 km) orbit, 15 000 lb (6 800 kg) into geostationary transfer orbit, 8 850 lb (4 014 kg) to Earth escape
Typical missions: Helios, Viking
Operational record: Latest Titan variant entered service in December 1974 with launch of Helios A
Prime contractor: Martin Marietta

AGENA

(USA — DOD/Nasa)

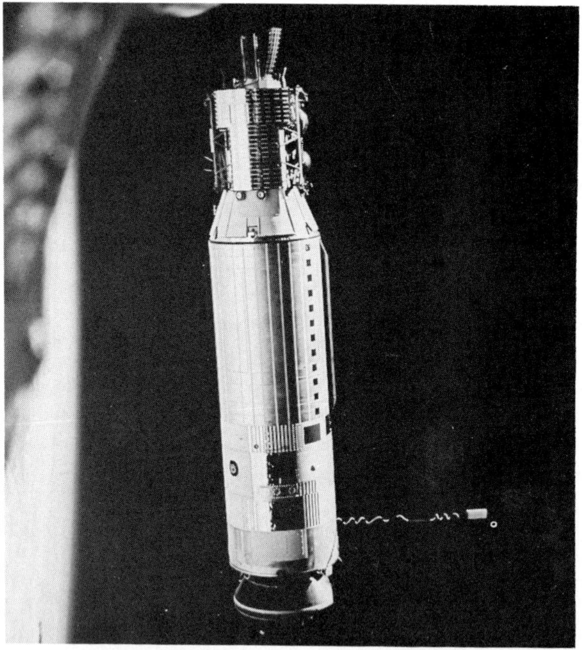

Data: Agena D
Configuration: Multi-purpose booster used as upper stage on modified versions of Thor IRBM and Atlas and Titan ICBMs. Also used by DOD as self-boosting satellite with instruments integrated (on earlier versions) or as attached package
Height: 22 ft (7.7 m)
Diameter: 5 ft (1.5 m)
Launch weight: 15 037 lb (6 821 kg)
Empty weight: 1 484 lb (673 kg)
Propulsion: Single Bell 8096 engine capable of multiple restarts. Propellants IRFNA/UDMH. Thrust 16 000 lb (71.2 kN)
Guidance and control: Inertial guidance using gas jets provides stage control alone
Performance: As per first stage
Operational record: Most extensively used US upper stage entered service with Thor in February 1959, with Atlas in February 1960, with Titan IIIB in July 1966. Total of at least 325 operational missions by end of 1974 includes 31 failures. Total includes about 47 payload launch applications (with two failures) and remainder as satellites
Prime contractor: Lockheed

CENTAUR

Data: Centaur D-1T
Configuration: Multi-purpose, high-energy booster used as upper stage for modified Atlas and Titan ICBMs. Latest Centaur version has D-1A and D-1T forms indicating first stage adaptation. D-1A has fitted extension payload shroud; D-1T is fully enclosed with payload inside Centaur Standard Shroud
Height: 30 ft (9.1 m). Shroud 58 ft (17.7 m)
Diameter: 10 ft (3.0 m). Shroud 14 ft (4.3 m)
Launch weight: 5 093 lb (2 310 kg) without shroud
Propulsion: Two Pratt & Whitney RL-10A engines capable of multiple restart. Propellants lox/liquid hydrogen. Thrust (vac) 15 000 lb (66.7 kN) each. Burn time (total) 563 sec. Isp 446 sec. Total of 14 hydrogen peroxide thrusters include eight of 3-50 lb (13-222 N) thrust for separation and propellant settling plus attitude control system of four 3.5 lb (15.6 N) and two 6 lb (26.7 N) jets
Guidance and control: Inertial guidance, supported by multi-purpose computor, controls complete launch vehicle
Performance: As per first stage
Operational record: Centaur was first US fully cryogenic engine and entered service with Atlas booster in May 1966. Total of 23 successes from 27 operational missions by end of 1974 (all failures due to first stage)
Prime contractor: General Dynamics Convair Division

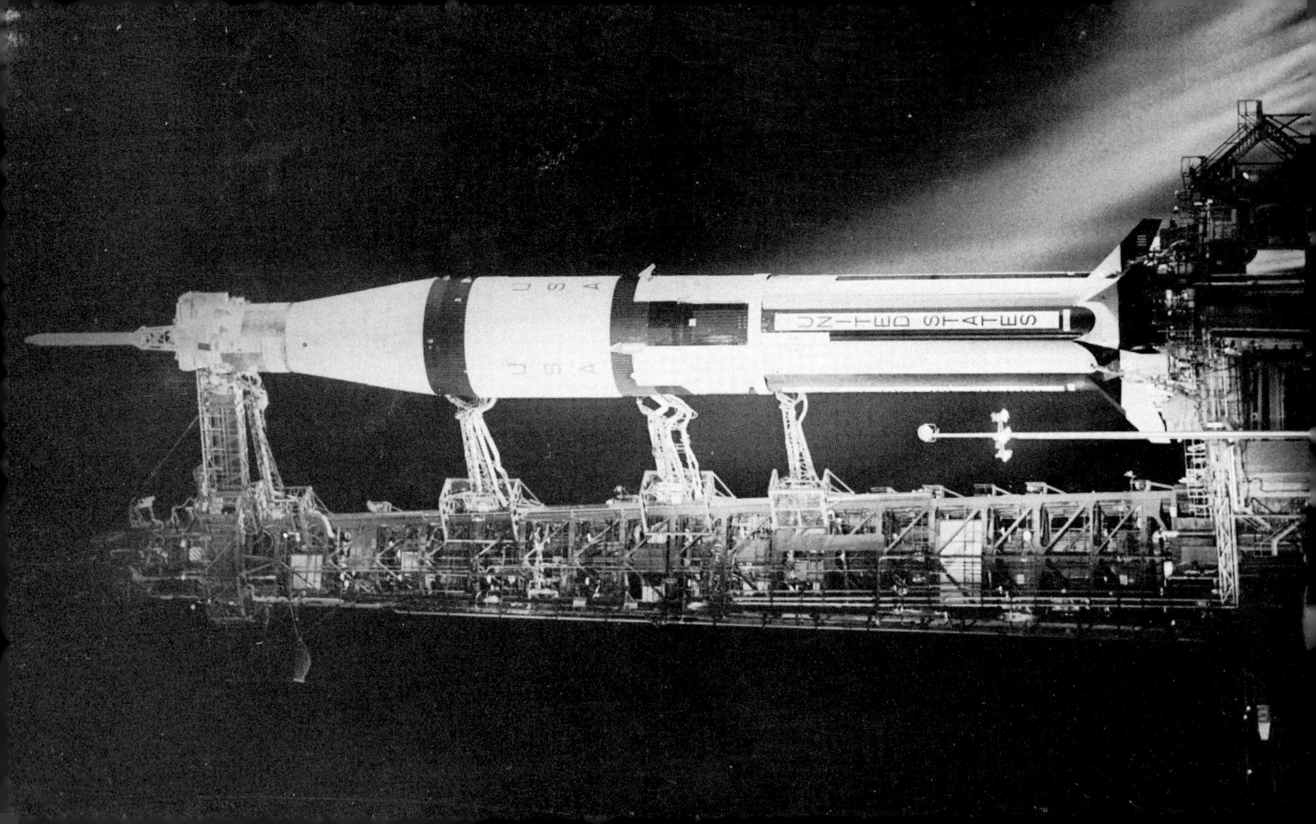
USA
UNITED STATES

Data: Saturn 1B
Configuration: Two-stage, multi-purpose, man-rated launch vehicle. First stage has cluster design with core lox tank surrounded by four lox and four fuel tanks
Height: 173 ft (56.2 m). With Apollo assembly 224 ft (68.2 m). First stage 80 ft 4 in (32.6 m). Second stage 58 ft 5 in (23.7 m). Instrument unit 3 ft (0.9 m). Shroud 31 ft 4 in (12.7 m)
Diameter: First stage 21 ft 5 in (6.5 m). Second stage 21 ft 8 in (6.6 m)
Launch weight: 1 300 000 lb (589 680 kg). First stage 882 000 lb (400 075 kg). Second stage 230 000 lb (104 328 kg)
Empty weight: 126 000 lb (57 244 kg) without payload/shroud. First stage 85 500 lb (38 782 kg). Second stage 24 800 lb (11 249 kg)
First stage propulsion: (S-1B) eight Rocketdyne H-1 engines; four outboard engines gimballed. Propellants lox/kerosene. Thrust 205 000 lb (911.8 kN) each. Burn time 125 sec
Second stage propulsion: (S-1VB) single Rocketdyne J-2 restartable engine. Propellants lox/liquid hydrogen. Thrust 215 000 lb (1000.8 kN). Burn time 294 sec
Guidance and control: Inertial guidance in Instrument unit uses first-stage gimballing and second-stage three-axis control-thrusters
Performance: 40 000 lb (18 144 kg) into 115 mile (185 km) orbit, 13 000 lb (5 897 kg) to Earth escape
Operational record: Total of eight launches between February 1966 and end of 1974 comprise one manned and four unmanned (two suborbital) development flights plus three Skylab crew launches. Last anticipated use was for ASTP mission in July 1975. No launch failures
Prime contractor: Chrysler (first stage), McDonnel Douglas (second stage)

USA

(USA — Nasa) SATURN V

Data: Saturn V (Apollo 17)
Configuration: Three-stage heavy launcher purpose-built for manned lunar missions
Height: 281 ft (85.6 m). With Apollo assembly 363 ft (110.6 m). First stage 138 ft (42.1 m). Second stage 81 ft 6 in (24.8 m). Third stage 59 ft 4 in (18.1 m). Instrument unit 3 ft (0.9 m)
Diameter: First/second stages 33 ft (10.1 m). Third stage/IU 21 ft 8 in (6.6 m)
Launch weight: 6 329 035 lb (2 870 850 kg) without payload/shrouds. First stage 4 952 755 lb (2 246 570 kg). Second stage 1 087 580 lb (493 318 kg). Third stage 266 000 lb (121 000 kg). IU 4 500 lb (2 040)
Empty weight: 415 594 lb (188 513 kg) without payload/shroud. First stage 287 574 lb (130 441 kg). Second stage 80 420 lb (36 478 kg). Third stage 24 900 lb (11 300 kg). IU 4 500 lb (2 040 kg)
First stage propulsion: (S-1C) five Rocketdyne F-1 engines. Outer four engines have gimballed nozzles. Propellants lox/kerosene. Thrust (s.1.) 1 533 022 lb (6 818.9 kN) each. Burn time 127 sec
Second stage propulsion: (S-II) five Rocketdyne J-2 engines with gimballed nozzles on four outer engines. Propellants lox/liquid hydrogen. Thrust 230 740 lb (1.026.3 kN). Burn time 365 sec
Third stage propulsion: (S-IVB) see Saturn 1B second stage
Guidance and control: Inertial guidance in instrument unit using gimballed nozzles for first two stages and thrusters for third. Emergency manual control from CM after first stage separation.
Performance: 230 000 lb (104 328 kg) into 115 mile (185 km) orbit, 95 000 lb (43 092 kg) to Earth escape
Operational record: Total of 13 launches between November 1967 and May 1973 comprise two unmanned launches, two lunar orbit, and one Earth orbit development flights, seven Apollo lunar landing missions (incl Apollo 13) and launch of Skylab. No launch failures. No further use anticipated
Prime contractor: Boeing (first stage), Rockwell (second stage), McDonnel Douglas (third stage)

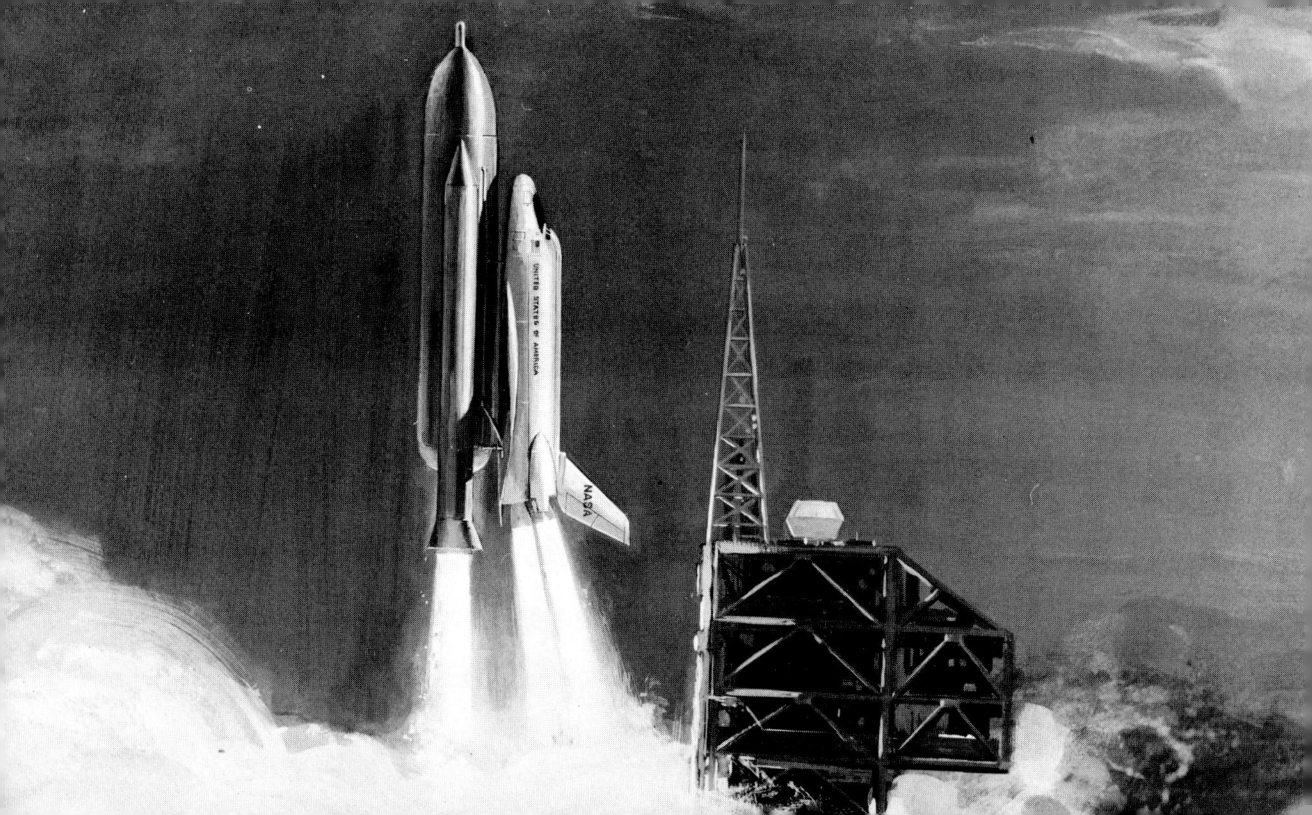
UNITED STATES OF AMERICA
NASA

SPACE SHUTTLE BOOSTER SYSTEM

Data: Space Shuttle prototype
Configuration: Two-stage parallel-burn booster system is man-rated launch facility for space shuttle orbiter. Comprises liquid-propellant space shuttle main engines (SSME) in orbiter, fed by external tank, plus two strap-on solid rocket boosters (SRB). All engines ignite at lift-off. SRBs jettison at 27 miles (43 km) altitude for recovery and re-use SSMEs continue as sustainer until near-orbit and tank-jettison
Length: Overall shuttle 184 ft (56.1 m). External tank 154 ft (46.9 m) SRB 149 ft (45.4 m)
Height: To orbiter fin-tip 76 ft (23.2 m)
Diameter: Orbiter wingspan 78 ft (23.8 m). External tank 27 ft 6 in (8.4 m). SRB 12 ft 2 in (3.7 m)
Launch weight: Overall shuttle 4.4 million lb (2.0 million kg). External tank 1 626 329 lb (737 721 kg). SRB (each) 1 281 160 lb (581 134 kg)
Empty weight: Overall shuttle 580 000 lb (263 000 kg). External tank 72 502 lb (32 932 kg). SRB (each) 179 160 lb (81 266 kg)
Orbiter propulsion: Three engines in cluster. 100-mission lifetime. Propellants lox/liquid hydrogen. Thrust (each — s.l.) 375 000 lb (1 688 kN), (vac) 475 000 lb (2 112.8 kN)
SRB propulsion: Two solid-fuel boosters, parachute recovered and re-usable up to ten times. Thrust (each — s.l.) 2.6 million lb (11 564.8 kN). Four separation thrusters at either end; thrust 15 000 lb (66.6 kN) each
Guidance and control: Orbiter inertial guidance using gimballed nozzles on SRBs, thrusters on orbiter
Performance: 230 000 lb (104 328 kg) into 115 mile (185 km) due-East orbit
Typical mission: Space shuttle orbiter
Operational record: Scheduled to enter service in 1980
Prime contractors: (SSME) Rocketdyne; (tank) Martin Marietta; (SRB) Thiokol

Appendix 1: Chronology of Manned Spaceflights

Name	Country	Launch Vehicle	Launch Date	Mission Duration Dys:Hr.Min	Crew	Mission
Vostok 1	USSR	A-1	April 12, 1961	0:01.48	Gagarin	Earth orbit; first manned spaceflight
Freedom 7 (MR3)	USA	Redstone	May 5, 1961	0:00.15	Shepard	Sub-orbital flight
Liberty Bell 7 (MR4)	USA	Redstone	July 21, 1961	0:00.15	Grissom	Sub-orbital flight
Vostok 2	USSR	A-1	August 6, 1961	1:01.16	Titov	Earth orbit
Friendship 7 (MA 6)	USA	Atlas	February 20, 1962	0:04.55	Glenn	First US orbital flight
Aurora 7 (MA 7)	USA	Atlas	May 24, 1962	0:04.56	Carpenter	Earth orbit
Vostok 3	USSR	A-1	August 11, 1962	3:22.22	Nikolayev	Earth orbit
Vostok 4	USSR	A-1	August 12, 1962	2:22.57	Popovich	Launched into co-orbit rendezvous with Vostok 3
Sigma 7 (MA 8)	USA	Atlas	October 3, 1962	0:09.13	Schirra	Earth orbit
Faith 7 (MA 9)	USA	Atlas	May 14, 1963	1:10.20	Cooper	Earth orbit
Vostok 5	USSR	A-1	June 14, 1963	4:23.6	Bykovsky	Co-orbit with Vostok 6
Vostok 6	USSR	A-1	June 16, 1963	2:22.50	Tereshkova	First woman in space
Voskhod 1	USSR	A-2	October 12, 1964	1:00.17	Komarov, Yegorov, Feoktistov	First multiple crew
Voskhod 2	USSR	A-2	March 18, 1965	1:02.02	Belyaev, Leonov	First extra-vehicular activity
Gemini 3	USA	Titan II	March 23, 1965	0:04.52	Grissom, Young	Earth orbit
Gemini 4	USA	Titan II	June 3, 1965	4:01.56	McDivitt, White	Earth orbit
Gemini 5	USA	Titan II	August 21, 1965	7:22.55	Cooper, Conrad	Earth orbit
Gemini 7	USA	Titan II	December 4, 1965	13:18.36	Borman, Lovell	Record flight; rendezvous target for Gemini 6
Gemini 6	USA	Titan II	December 15, 1965	1:01.51	Schirra, Stafford	First manoeuvred rendezvous
Gemini 8	USA	Titan II	March 16, 1966	0:10.41	Armstrong, Scott	First docking, using Agena target
Gemini 9	USA	Titan II	January 6, 1966	3:00.21	Stafford, Cernan	First US EVA
Gemini 10	USA	Titan II	July 18, 1966	2:22.47	Young, Collins	Earth orbit
Gemini 11	USA	Titan II	September 12, 1966	2:23.17	Conrad, Gordon	Earth orbit
Gemini 12	USA	Titan II	November 11, 1966	3:22.35	Lovell, Aldrin	Earth orbit
Apollo 4	USA	Saturn 1B	January 27, 1967	—	Grissom, White, Chaffee	Crew killed in launch-pad fire
Soyuz 1	USSR	A-2	April 23, 1967	2:02.40	Komarov	Crewman killed on landing
Apollo 7	USA	Saturn 1B	October 11, 1968	10:20.09	Schirra, Cunningham, Eisele	First Apollo test
Soyuz 3	USSR	A-2	October 26, 1968	3:22.51	Beregovoy	Rendezvous with unmanned Soyuz 2
Apollo 8	USA	Saturn V	December 21, 1968	6:03.01	Borman, Lovell, Anders	First manned circumlunar flight
Soyuz 4	USSR	A-2	January 14, 1969	2:23.14	Shatalov	First USSR docking, first crew transfer
Soyuz 5	USSR	A-2	January 15, 1969	3:00.46	Volynov. Khrunov, Yeliseyev	Khrunov, Yeliseyev transfer by EVA
Apollo 9	USA	Saturn V	March 3, 1969	10:01.01	McDivitt, Scott, Schweickart	Earth orbit test of Lunar Module

Name	Country	Launch Vehicle	Launch Date	Mission Duration Dys:Hr.Min	Crew	Mission
Apollo 10	USA	Saturn V	May 18,1969	8:00.03	Stafford, Cernan, Young	Lunar orbit test of Lunar Module
Apollo 11	USA	Saturn V	July 16, 1969	8:03.19	Armstrong, Aldrin, Collins	First Moon landing on July 20
Soyuz 6	USSR	A-2	October 11, 1969	4:22.42	Shonin, Kubasov	First of three craft in co-orbit link-up
Soyuz 7	USSR	A-2	October 12, 1969	4:22.41	Filipchenko, Volkov, Gorbatko	Earth orbit
Soyuz 8	USSR	A-2	October 13, 1969	4:22.50	Shatalov, Yeliseyev	Moon landing
Apollo 12	USA	Saturn V	November 14, 1969	10:04.36	Conrad, Bean, Gordon	Moon landing
Apollo 13	USA	Saturn V	April 11, 1970	5:22.55	Lovell, Haise, Swigert	Landing cancelled after en-route explosion
Soyuz 9	USSR	A-2	June 1, 1970	17:16.59	Nikolayev, Sevastyanov	Long duration medical test
Apollo 14	USA	Saturn V	January 31, 1971	9:00.02	Shepard, Mitchell, Roosa	Moon landing
Soyuz 10	USSR	A-2	April 22, 1971	1:23.46	Shatalov, Yeliseyev, Rukavishnikov	Docked with Salyut 1
Soyuz 11	USSR	A-2	June 6, 1971	23:18.21	Dobrovolski, Volkov, Patsayev	First Salyut crew killed during re-entry
Apollo 15	USA	Saturn V	July 26, 1971	12:07.12	Scott, Irwin, Worden	First Lunar Roving vehicle
Apollo 16	USA	Saturn V	April 16, 1972	11:01.51	Young, Duke, Mattingly	Moon landing
Apollo 17	USA	Saturn V	December 7, 1972	12:13.51	Cernan, Schmidt, Evans	Final Moon landing
Skylab 2	USA	Saturn 1B	May 25, 1973	28:00.49	Conrad, Kerwin, Weitz	First Skylab crew
Skylab 3	USA	Saturn 1B	July 28, 1973	59:11.09	Bean, Lousma, Garriot	Second Skylab crew
Soyuz 12	USSR	A-2	September 27, 1973	1:23.16	Lazarev, Makarov	Test of modifications
Skylab 4	USA	Saturn 1B	November 16, 1973	84:01.16	Carr, Gibson, Pogue	Final Skylab crew
Soyuz 13	USSR	A2	December 18, 1973	7:20.50	Klimuk, Lebedev	ASTP rehearsal
Soyuz 14	USSR	A-2	July 3, 1974	15:17.30	Popovitch, Artyukhin	Salyut 3 crew
Soyuz 15	USSR	A-2	August 26, 1974	2:00.12	Sarafanov, Demin	Salyut 3 crew failed to dock
Soyuz 16	USSR	A-2	December 12, 1974	2:22.24	Filipchenko, Rukavshnikov	ASTP rehearsel
Soyuz 17	USSR	A-2	January 10, 1975	29:13.20	Gubarev, Grechko	Salyut 4 crew
Soyuz 18A	USSR	A-2	April 5, 1975	—	Lazarev, Makarov	Launch failure; crew escaped
Soyuz 18	USSR	A-2	May 24, 1975	62:23.20	Klimik, Sevastyanov	Salyut 4 crew
Soyuz 19	USSR	A-2	July 15, 1975	5:22.31	Leonov, Kubasov	ASTP flight
Apollo	USA	Saturn 1B	July 15, 1975	9:02.28	Stafford, Brand, Slayton	ASTP flight

Appendix 2: Chronology of interplanetary probes

Name	Country	Launch Vehicle	Launch date	Launch site	Objective	Mission result
Pioneer 0	USA	Thor Able 1	August 17,1958	Cape Canaveral	Lunar orbit	Launch failure
Pioneer 1	USA	Thor Able 1	October 11, 1958	Cape Canaveral	Lunar orbit	Launch failure
Pioneer 2	USA	Thor Able 1	November 8, 1958	Cape Canaveral	Lunar orbit	Launch failure
Pioneer 3	USA	Juno II	December 6, 1958	Cape Canaveral	Lunar impact	Launch failure
Luna 1	USSR	A.1	January 2, 1959	Tyuratam	Lunar impact	Missed Moon
Pioneer 4	USA	Juno II	March 3, 1959	Cape Canaveral	Lunar impact	Missed Moon
Lunar 12	USSR	A.1	September 12, 1959	Tyuratam	Lunar impact	First lunar impact
Pioneer (P-1)	USA	Atlas able	September 24, 1959	Cape Canaveral	Lunar orbit	Launcher exploded on pad
Luna 3	USSR	A.1	October 4, 1959	Tyuratum	Lunar fly-by	First farside photography
Pioneer (P-3)	USA	Atlas Able	November 26, 1959	Cape Canaveral	Lunar orbit	Launch failure
Pioneer (P-30)	USA	Atlas Able	September 25, 1960	Cape Canaveral	Lunar orbit	Launch failure
U-1	USSR	A-2-e	October 10, 1960	Tyuratam	Mars	Failed to leave Earth orbit
U-2*	USSR	A-2-e	October 14, 1960	Tyuratam	Mars	Failed to leave Earth orbit
Pioneer (P-31)	USA	Atlas Able	December 15, 1960	Cape Canaveral	Lunar orbit	Launch failure
Tyazhyoliy Sputnik 4	USSR	A-2-e	February 4, 1961	Tyuratam	Venus	Failed to leave Earth orbit
Venera 1	USSR	A-2-e	February 12, 1961	Tyuratam	Venus fly-by	Communications lost en-route
Ranger 3	USA	Atlas Agena B	January 26, 1962	Cape Canaveral	Lunar impact	Missed Moon
Ranger 4	USA	Atlas Agena B	April 23, 1962	Cape Canaveral	Lunar impact	Spacecraft failure
Mariner 1	USA	Atlas Agena B	August 22, 1962	Cape Canaveral	Venus fly-by	Launch failure
U-3*	USSR	A-2-e	August 25, 1962	Tyuratam	Venus	Failed to leave Earth orbit
Mariner 2	USA	Atlas Agena B	August 27, 1962	Cape Canaveral	Venus fly-by	First planetary fly-by
U-4*	USSR	A-2-e	September 1, 1962	Tyuratam	Venus	Failed to leave Earth orbit
U-5*	USSR	A-2-e	September 12, 1962	Tyuratam	Venus	Failed to leave Earth orbit
Ranger 5	USA	Atlas Agena B	October 18, 1962	Cape Canaveral	Lunar impact	Missed Moon
U-6*	USSR	A-2-e	October 24, 1962	Tyuratam	Mars	Exploded in Earth orbit
Mars 1	USSR	A-2-e	November 1, 1962	Tyuratam	Mars	Communications lost-en-route
U-7*	USSR	A-2-e	November 4, 1962	Tyuratam	Mars	Failed to leave Earth orbit
U-8*	USSR	A-2-e	January, 4, 1963	Tyuratam	Lunar lander	Failed to leave Earth orbit
Luna 4	USSR	A-2-e	April 2, 1963	Tyuratam	Lunar lander	Missed Moon
Cosmos 21	USSR	A-2-e	November 11, 1963	Tyuratam	Venus	Engineering test; remained in Earth orbit
Ranger 6	USA	Atlas Agena B	January 30, 1964	Kennedy Space Center	Lunar impact/TV	Impact but cameras failed
Cosmos 27	USSR	A-2-e	March 27, 1964	Tyuratam	Venus	Failed to leave Earth orbit

*Unannounced mission identified by US sources

Name	Country	Launch Vehicle	Launch date	Launch site	Objective	Mission result
Zond 1	USSR	A-2-e	April 2, 1964	Tyuratam	Venus	Communications lost en-route
Ranger 7	USA	Atlas Agena D	July 28, 1964	Kennedy Space Center	Lunar impact/TV	Successful approach photography
Mariner 3	USA	Atlas Agena D	November 5, 1964	Kennedy Space Center	Mars fly-by	Launch failure
Mariner 4	USA	Atlas Agena D	November 28, 1964	Kennedy Space Center	Mars fly-by	Successful fly-by photography
Zond 2	USSR	A-2-e	November 30, 1964	Tyuratam	Mars fly-by	Communications lost en-route
Ranger 8	USA	Atlas Agena B	February 17, 1965	Kennedy Space Center	Lunar impact/TV	Success
Cosmos 60	USSR	A-2-e	March 12, 1965	Tyuratam	Lunar landing	Failed to leave Earth orbit
Ranger 9	USA	Atlas Agena B	March 21, 1965	Kennedy Space Center	Lunar impact/TV	Success
Luna 5	USSR	A-2-e	May 9, 1965	Tyuratam	Lunar landing	Crashed on moon
Zond 3	USSR	A-2-e	July 18, 1965	Tyuratam	Mars fly-by test	Engineering flight; passed Mars orbit
Venera 2	USSR	A-2-e	November 12, 1965	Tyuratam	Venus	Communications lost en-route
Venera 3	USSR	A-2-e	November 16, 1965	Tyuratam	Venus impact	First Venus impact
Cosmos 96	USSR	A-2-e	November 23, 1965	Tyuratam	Venus	Failed to leave Earth orbit
Pioneer 6	USA	T-A Delta	December 16, 1965	Kennedy Space Center	Solar orbit	Success; first solar monitor
Luna 9	USSR	A-2-e	January 31, 1966	Tyuratam	Lunar landing	First lunar soft landing
Cosmos 111	USSR	A-2-e	March 1, 1966	Tyuratam	Lunar orbit	Failed to leave Earth orbit
Luna 10	USSR	A-2-e	March 31, 1966	Tyuratam	Lunar orbit	First lunar orbiter
Surveyor 1	USA	Atlas Centaur	May 30, 1966	Kennedy Space Center	Lunar landing	Success
Explorer 33	USA	T-A Delta	July 1, 1966	Kennedy Space Center	Lunar orbit	Failed to achieve orbital capture
Lunar Orbiter 1	USA	Atlas Agena D	August 10, 1966	Kennedy Space Center	Lunar orbit	Success
Pioneer 7	USA	T-A Delta	August 17, 1966	Kennedy Space Center	Solar orbit	Success
Luna 11	USSR	A-2-e	August 24, 1966	Tyuratam	Lunar orbit	Success; no photography
Surveyor 2	USA	Atlas Centaur	September 24, 1966	Kennedy Space Center	Lunar landing	Crashed on Moon
Luna 12	USSR	A-2-e	October 22, 1966	Tyuratam	Lunar orbit	Success
Lunar Orbitor 2	USA	Atlas Agena D	November 6, 1966	Kennedy Space Center	Lunar orbit	Success
Luna 13	USSR	A-2-e	December 21, 1966	Tyuratam	Lunar landing	Success
Lunar Orbiter 3	USA	Atlas Agena D	February 4, 1967	Kennedy Space Center	Lunar orbit	Success
Surveyor 3	USA	Atlas Centaur	April 17, 1967	Kennedy Space Center	Lunar landing	Success; including soil test
Lunar Orbiter 4	USA	Atlas Agena D	May 4, 1967	Kennedy Space Center	Lunar orbit	Success
Venera 4	USSR	A-2-e	June 12, 1967	Tyuratam	Venus impact	Penetrated atmosphere
Mariner 5	USA	Atlas Agena D	June 14, 1967	Kennedy Space Center	Venus fly-by	Success
Cosmos 167	USSR	A-2-e	June 17, 1967	Tyuratam	Venus	Failed to leave Earth orbit
Surveyor 4	USA	Atlas Centaur	July 14, 1967	Kennedy Space Center	Lunar landing	Communications lost before touchdown

Name	Country	Launch Vehicle	Launch date	Launch site	Objective	Mission result
Explorer 35	USA	T-A Delta	July 19, 1967	Kennedy Space Center	Lunar orbit	Success
Lunar Orbiter 5	USA	Atlas Agena D	August 2, 1967	Kennedy Space Center	Lunar orbit	Success
Surveyor 5	USA	Atlas Centaur	September 8, 1967	Kennedy Space Center	Lunar landing	Success; first soil analysis
Surveyor 6	USA	Atlas Centaur	November 7, 1967	Kennedy Space Center	Lunar landing	Success
Pioneer 8	USA	T-A Delta	December 13, 1967	Kennedy Space Center	Solar orbit	Success
Surveyor 7	USA	Atlas·Centaur	January 7, 1968	Kennedy Space Center	Lunar landing	Success
Zond 4	USSR	A-2-e	March 2, 1968	Tyuratam	Circumlunar	Engineering test to orbit of Moon
Luna 14	USSR	A-2-e	April 7, 1968	Tyuratam	Lunar orbit	Success
Zond 5	USSR	A-2-e	September 14, 1968	Tyuratam	Circumlunar	Success; first interplanetary recovery
Pioneer 9	USA	T-A Delta	November 8, 1968	Kennedy Space Center	Solar orbit	Success
Zond 6	USSR	A-2-e	November 10, 1968	Tyuratam	Circumlunar	Success
Venera 5	USSR	A-2-e	January 5, 1969	Tyuratam	Venus impact	Success
Venera 6	USSR	A-2-e	January 10, 1969	Tyuratam	Venus impact	Success
Mariner 6	USA	Atlas Centaur	February 24, 1969	Kennedy Space Center	Mars fly-by	First Mars photography
Mariner 7	USA	Atlas Centaur	March 27, 1969	Kennedy Space Center	Mars fly-by	Success
Luna 15	USSR	D-1-e	July 13, 1969	Tyuratam	Lunar landing	Crashed in landing attempt
Zond 7	USSR	D-1-e	August 7, 1969	Tyuratam	Circumlunar	Success
Pioneer E	USA	LTTA Delta	August 27, 1969	Kennedy Space Center	Solar orbit	Launch failure
Cosmos 300	USSR	D-1-e	September 23, 1969	Tyuratam	Lunar landing?	Failed to leave Earth orbit
Cosmos 305	USSR	D-1-e	October 22, 1969	Tyuratam	Lunar landing?	Failed to leave Earth orbit
Venera 7	USSR	A-2-e	August 17, 1970	Tyuratam	Venus landing	First signals from surface
Cosmos 359	USSR	A-2-e	August 22, 1970	Tyuratam	Venus	Failed to leave Earth orbit
Luna 16	USSR	D-1-e	September 22, 1970	Tyuratam	Lunar landing	First automatic sample collector
Zond 8	USSR	D-1-e	October 20, 1970	Tyuratam	Circumlunar	Success
Luna 17	USSR	D-1-e	November 10, 1970	Tyuratam	Lunar landing	First Lunokhod rover
Mariner 8	USA	Atlas Centaur	May 8, 1971	Kennedy Space Center	Mars orbit	Launch failure
Cosmos 419	USSR	D-1-e	May 10, 1971	Tyuratam	Mars	Failed to leave Earth orbit
Mars 2	USSR	D-1-e	May 19, 1971	Tyuratam	Mars orbit/landing	Suceeded but lander impacted
Mars 3	USSR	D-1-e	May 28, 1971	Tyuratam	Mars orbit/landing	Capsule achieved first soft landing.
Mariner 9	USA	Atlas Centaur	May 30, 1971	Kennedy Space Center	Mars orbit	Success
Luna18	USSR	D-1-e	September 2, 1971	Tyuratam	Lunar landing	Communications lost at touch down
Luna 19	USSR	D-1-e	September 28, 1971	Tyuratam	Lunar orbit	Success
Luna 20	USSR	D-1-e	February 25, 1972	Tyuratam	Lunar landing	Second automatic sample collector

Name	Country	Launch Vehicle	Launch date	Launch site	Objective	Mission result
Pioneer 10	USA	Atlas Centaur	March 3, 1972	Kennedy Space Center	Jupiter fly-by	Success; first close-up pictures
Venera 8	USSR	A-2-e	March 27, 1972	Tyuratam	Venus landing	First Sun-side landing
Cosmos 482	USSR	A-2-e	March 31, 1972	Tyuratam	Venus landing	Failed to leave Earth orbit
Luna 21	USSR	D-1-e	January 8, 1973	Tyuratam	Lunar landing	Second Lunokhod rover
Pioneer 11	USA	Atlas Centaur	April 5, 1973	Kennedy Space Center	Jupiter fly-by	En-route for Saturn after swing-by
Explorer 49	USA	Delta	June 10, 1973	Kennedy Space Center	Lunar orbit	Success
Mars 4	USSR	D-1-e	July 21, 1973	Tyuratam	Mars orbit	Failed to achieve orbital capture
Mars 5	USSR	D-1-e	July 25, 1973	Tyuratam	Mars orbit	Success
Mars 6	USSR	D-1-e	August 5, 1973	Tyuratam	Mars landing	Lander communications lost before touchdown
Mars 7	USSR	D-1-e	August 9, 1973	Tyuratam	Mars Landing	Lander missed planet after release
Mariner 10	USA	Atlas Centaur	November 3, 1973	Kennedy Space Center	Mercury fly-by	First Mercury flight; also flew by Venus
Luna 22	USSR	D-1-e	March 21, 1974	Tyuratam	Lunar orbit	Success
Luna 23	USSR	D-1-e	October 28, 1974	Tyuratam	Lunar landing	Soil collector made crash landing
Venera 9	USSR	D-1-e	June 8, 1975	Tyuratam	Venus landing	First surface photographs
Venera 10	USSR	D-1-e	June 14, 1975	Tyuratam	Venus landing	Successful repeat
Viking 1	USA	Titan Centaur	August 10, 1975	Kennedy Space Center	Mars landing	
Viking 2	USA	Titan Centaur	September 9, 1975	Kennedy Space Center	Mars landing	

INDEX